BLACK CLIMATES

Black Climates

Notes on Race, Our Environment and Visions for Equitable Futures

SELINA NWULU

Chatto & Windus
LONDON

1 3 5 7 9 10 8 6 4 2

Chatto & Windus, an imprint of Vintage, is part of the
Penguin Random House group of companies

Vintage, Penguin Random House UK, One Embassy Gardens,
8 Viaduct Gardens, London SW11 7BW

penguin.co.uk/vintage
global.penguinrandomhouse.com

First published by Chatto & Windus in 2025

Typeset in 12/14.75 pt Dante MT Pro by Six Red Marbles UK, Thetford, Norfolk
Printed and bound in Great Britain by Clays Ltd, Elcograf S.p.A.

The authorised representative in the EEA is Penguin Random House Ireland,
Morrison Chambers, 32 Nassau Street, Dublin D02 YH68

A CIP catalogue record for this book is available from the British Library

ISBN 9781784744885

Penguin Random House is committed to a sustainable future
for our business, our readers and our planet. This book is made
from Forest Stewardship Council® certified paper.

For the tender-hearted, broken by the state of the world. For those committed to unseen, often silent, acts of bravery, willing to stand within what is difficult and unknown. To those who choose to live in vibration with a better world, who are stewarding its arrival.

For you.

Contents

Contents

Contents

Contents

Letter to the Reader

Dear reader,

I've tried to write to you many times.

Writing to you in a world undergoing rapid change has too often felt at odds with the permanence that a book offers.

Week after week, there's another crisis: unpredictable seasons and extreme weather conditions, political unrest, an ongoing cost-of-living crisis and creeping extremism. In the span of writing this book, there's been a dizzying number of government scandals, and a change of government – not just in the UK, but also across Europe and in the United States – with many of those same countries continuing to provide economic support to regimes accused of genocides worldwide.

This book has come into existence amid the waning and emergence of global powers, an ongoing pandemic and an intensifying threat of global conflict and war. It's clear that we're in a very troubled period in our collective history. And any attempt to make sense of what we see around us is being continually undermined by the shifting goalposts of our time. I am writing to you amid turbulence – of systems collapsing in on themselves – deep societal changes that will shape us, one way or another, into something else.

However, what remains unchanged is the existential challenge posed by the climate crisis, which threatens the quality of our existence on a collective level. Of course, its impact is never felt all at once, but rather in different ways – in moments fast and slow. This is a crisis that splits time and experience. While a future world of climate catastrophe may be a distant concern for some, it's already a lived reality for the most vulnerable and disadvantaged among us. Just like any crisis, climate change will continue to disproportionately affect the most marginalised first. Our exploration of this in Black Climates situates these ideas within our complicated and changeable realities, putting what – and who – has been historically overlooked and minimised around the climate crisis front and centre in these pages.

The ongoing difficulties over the past few years have also left me wondering about the violence we inflict on one another as a consequence of surviving under oppressive state powers. Beyond any topic the book covers is the central question of how we overcome this harm and disconnection and find our way back to ourselves and each other.

So often when we think about the climate crisis, we think about the inaction of state and corporate powers – and that's entirely correct. It's up to our governments to shoulder the burden of the crisis, given that many of their decisions have led us to this point. Rest assured, there will be no talk of us needing to recycle more or use paper straws here. But in coming to know this book, as I hope you will, Black Climates is as much – if not more – about a personal revolution, one that transcends anything we're waiting to happen outside of ourselves, and one that's entirely of our own courageous making.

In love and solidarity,
Selina

Introduction

As in a disaster dream, catastrophe piles on catastrophe

Richard Seymour[1]

This is not just another crisis – this is a breaking point.

We're living through the consequences of generations of harm to the planet and its people, driven by the hand of colonialism and ongoing exploitation.

If there's one thing I've learned, it's that all living things seek equilibrium. The earth seeks to achieve balance by any means.

There's no shadow without light, no loud without quiet, no bitter without sweet.

The climate crisis is a message about which systems and ways of being cannot continue – a warning sign that the planet cannot be outrun. It demands we change.

What's out of balance in your life and what are you willing to let go?

One thing that's consistent in conversations I've had with people about climate change over the years is a nervousness. When I mention my writing or any research I've done around the topic, it's clear that it's a Very Important Subject. But beyond that, the climate crisis can often feel unclear – a hard subject without many accessible inroads. And let's be honest, it can also be a pretty dry issue to engage with. This is because, for way too long, the climate crisis has been a topic dominated by the academic and scientific community, all worrying projections and graphs, but lacking in stories that speak to the majority.

The mainstream climate movement has also historically had us believe that climate change is yet another thing we should worry about, beyond the day-to-day oppression of Black lives. This is deeply inaccurate, especially since the root cause of the crisis has always been closely tied to historical colonial powers. The exploitation of resource-rich countries across the globe led to rapid industrial expansion via the extraction of more and more resources, which in turn polluted and decimated natural ecosystems all over the colonised world. While European powers were 'scrambling for Africa' and creating artificial borders for an entire continent, they were simultaneously devastating entire ecologies, communities and local knowledge on how to tend and maintain the land.[2]

Today we live within this legacy – not least through the many families of the Black diaspora who have left their countries of origin to settle elsewhere. I don't blame anyone with this as part of their history for distancing themselves from climate change discussions; if someone else made a mess in your back garden without your permission, why would you want to clean it up?

But how we hold and frame these conversations is vital. Because the reality is Black people are globally among the most affected by climate change. It's a trap to buy into false narratives that gloss over the realities of the crisis, or that claim that the bad things are happening elsewhere, as is so often the implicit assumption in the UK. For Black communities, 'elsewhere' is all too often in the parts of the world where we and our families live – whether that's within our cities, towns or countries that disproportionately bear the brunt of the worst effects of climate change. The crisis isn't yet another issue for us to get our heads around beyond the racisms we are currently facing – climate change is itself fundamentally racist. Recognising the intersections between race and the climate crisis offers us an expanded framework through which we can better understand and resist our oppression – it does not create a new one.

If racism is an ongoing attempt to exclude and undermine our humanity, why would mainstream conversations around climate change have ever been any different? Our exclusion within the climate movement exists within the same tired structure of racism that at best ignores us and at its worst exploits and destroys us. In the same vein, the messages of the climate movement – the talk about the problems and threats of crisis – simply put, were never talking to us, which is why so many of us have felt disinterested and disengaged.

The work of many climate activists to bring race into the climate conversation has seen increased mainstream understanding about the crisis and who it will affect the most. But this is still relatively new. I remember, not that long ago, being in meetings and talks where climate justice – a framework which centres the rights of people and groups most affected by the crisis – was a fringe idea. While the climate movement is beginning to head in a more intersectional direction, it still must reckon with the ways it has largely ignored our struggles within its mission. We also have to continue the work of making the overlooked connections between racism and climate change for ourselves, as is crucial to our survival.

Climate change speaks to how human activity is causing the planet to heat up in ways that are unsustainable for planetary life as we know it. While the planet has always been undergoing some sort of climatic change, it's the intense human activity since the 1800s that's been speeding up the process.[3] This, in large part, is caused by the release of greenhouse gases – such as carbon dioxide (CO_2) – into the atmosphere as a by-product of burning fossil fuels such as coal, gas and oil to produce energy. These greenhouse gases create a layer around the earth which traps the sun's heat.[4]

The knock-on effects of this include rising temperatures, food insecurity, melting glaciers and rising sea levels, which threaten to flood many cities and towns worldwide, not to mention a host of natural disasters we're ill-prepared for.[5]

Consider this the school assembly definition of climate change. But for me, beyond this, the climate crisis is a living and messy

6

topic, constantly interacting with the complexities of our lives. It's the way the conditions of our planet are changing, yes – but it's also about our immediate environment and the quality of the lives we live. It's about the food in our fridge, the places we live and work, the quality of the air we breathe, the things we value on a personal and collective level – and how we build community. It's in our connection to – and disconnection from – the world around us.

All this is underpinned by capitalism, a system at the core of which lie wealth and profit. Central to capitalism is private ownership – where companies and individuals create and operate businesses for profit. Straightforward enough, some might say, but it's a structure that has cast a permanent shadow over all our lives, prioritising everything based on its economic value, including the planet. In a world under the spell of capitalism, the economy reigns supreme, and all alternative suggestions for the good of the world – better though they may be – must bow to preserve the precious economy, and its ability to keep the capitalist wheels turning.

This inextinguishable drive for profit at all costs is at the heart of the climate crisis, on a planet with abundance for everyone, yet never enough under capitalism. Instead, essentials we all need to survive like food, warmth and shelter come with a price tag that fluctuates at the whim of the market. Things that should never be sold, have been. This includes us, in big ways like slavery, but also in seemingly smaller, more routine ways, like how most of us have to sacrifice our health, well-being and relationships in order to earn money and survive. Under capitalism, we're merely products to be separated from one another and exploited, much like the land and ecosystems across the world that have been stolen and plundered.

Late-stage capitalism, our current reality, is where we're surviving the very extremes of the system, of hyper-consumption and extreme cruelties. Everything – including our time and attention – is up for sale, and wealth inequalities loom large.

The climate crisis, as I outline in the book, invites us to return to a sense of interconnectedness, of understanding that the way we love and care for ourselves mirrors how we love and care for the planet – it's in everything. The more I think about it, the more I believe this to be true: we are the mirror image of the world we live in. And the destruction and devastation of the planet reflect the toxicity of the architects of this exploitation – the imperial powers and multinational corporations who are causing great harms at our expense.

The different essays in this book offer some clarity on what's at stake for Black people, and how the climate crisis connects to many aspects of our daily lives. We will explore this in five parts: Air, Borders, Land, How We Survive and How We Thrive, with both beginning and end focusing on things that can't be touched – though in very different ways. Some parts won't feel overtly like a 'climate change' book – but that's the point. The crisis is always there – whether implicit or explicit – and we need newer, more interwoven ways of understanding this, particularly through a Black lens. Climate change is as much about the damage to the land as it is about the ripple effects on our communities, bodies, spirits and minds.

Each essay in the book consists of several sections, each exploring the essay's themes from a different perspective. This allows for a more accessible breakdown of each topic while providing space to pause and ground after some of the heavier subject areas. In reflecting on how we reconnect with ourselves and the planet, it's also clear to me that beneath the different topics I cover – from air pollution to disability and land – lies the same story: what has been taken from us and the question of how we reclaim it. Just as land was divided under imperial rule, so too were our experiences, ways of living and understandings of ourselves and one another. This book attempts to offer a fuller, interconnected picture.

There is no understanding of the climate crisis without understanding ourselves as part of the ecosystems we exist in that are integral to the planet. Human nature is still nature. In the context of deep pain and societal rejection of the Black body, the planet wordlessly offers us a belonging that has rarely been afforded to us elsewhere. I hope this duality of holding both the pain and potential for deep belonging on this planet in a more embodied way reignites our participation, as the climate crisis pushes us into a different chapter in our collective world history.

For those already well versed in the realities of the climate crisis, I hope this offers a different articulation of how we can understand and speak about climate change. But my real hope is that the book brings more people who have been disengaged from conversations about the climate crisis into the fold. Much of what can be done to mitigate the crisis is beyond our control, but I believe widening the conversation means more public pressure to hold those in power to account. More importantly, my hope is that it also provides greater impetus for organising and community building among ourselves.

All we have is us.

A note on who I write this for:

I write these essays with Black people and communities in mind, particularly those of the diaspora – a community that straddles multiple cultures, often historically and presently in tension with one another. It's a warring and fragmented place to exist in, and one that, as a British Nigerian, I also inhabit. This is who the 'we' and 'us' within this book refers to. While I hope this isn't the sole audience that finds this work, this is who I primarily write it for. I like the way professor and author of *Black Disability Politics*, Sami Schalk, frames this when she says:

> If you are a non-Black person, imagine that you have just walked past my living room, where I'm having a conversation with my Black family and friends. You're welcome to come in, listen and learn from the conversation, maybe even contribute to it when appropriate, but the conversation, the space of this book, is not for or about you.[6]

The discourse on race has noticeably expanded in recent years, yet much of it still centres whiteness. I'm acutely aware of how often these conversations are happening *about* us, but not *with* us. While the country suffers from a self-imposed form of arrested development – always needing to question the same basics of racism and Black existence – there is little room left for our growth. We are all too often trapped in a never-ending loop, always having the same conversations and defending our existence rather than expanding upon it. This is what Toni Morrison means when she describes racism as a distraction:

'It keeps you from doing your work,' she says. 'It keeps you explaining, over and over again, your reason for being . . . None of this is necessary. There will always be one more thing.'[7]

How much space are we afforded for our growth? If the climate crisis demands we change and evolve, how would we like to do this? How often can we speak about our concerns, not just in a way that is defending ourselves, which is so often the case when Blackness is put in opposition to whiteness? Who are we beyond racism? Most of us know how to talk about white supremacy as a matter of survival – but at what cost? Do we know ourselves just as well? It's these truths and areas of exploration that we're being distracted from. This book is about how we evolve and define the parameters of our experiences for ourselves.

For too long, we've had to work within a troubling framework – one in which the global majority can be lumped together and with the assumption that, by doing so, something meaningful can be understood about the nuances of our cultures, experiences and lives. White people make up around just 15 per cent of the world's population,[8] which makes the misguided belief that our non-whiteness is our most common identity a ridiculous idea – one that is itself a product of white supremacist logic. My intention is to move past the 'BAME-ifaction'* of non-white communities and broad-brush ways of thinking about us that are unhelpful and overly generalised. Therefore, other marginalised racial communities won't be the focus of this book either. While our experiences may overlap in ways that will resonate throughout these pages, we also have many distinct realities that deserve their own space.

My approach to this is, of course, messy and imperfect because trying to untangle ourselves from these ways of thinking is lifetime work. I make no attempt to smooth over these complexities, because the struggle to articulate ourselves beyond imposed acronyms reflects legacies of great erasure and harm, which produced them in the first place. My interviews will also draw from the expertise of people of many ethnicities, and I will bring in the experiences of other racialised communities, as and when it feels relevant to do so, because the goal is never to erase.

My driving force for these essays is to adopt a stance of specificity in the Black experience. By doing so, I hope to create space for both the similarities and the differences within our experiences, beyond the blanket term of 'Blackness'.

Space has been an unexpected, recurring theme of this book – the spaces we live in, and those we're pushed out of or into. So, my intention is to offer us room to move past the rigidity of

---

* 'BAME' is an acronym which stands for Black, Asian, and Minority Ethnic. It was developed as a policy framework to capture data of the non-white population. However, given its overly broad and reductive nature, its usage is – thankfully – declining in popularity.

Blackness when it's reduced to an acronym or defined solely as a contrast to whiteness. It's in this way that we might see not only the depths to which we continue to be oppressed, but also the ways in which we, too, can harm and oppress others.

When given space – beyond the state of flight or fight that our oppression often pushes us into – how might we hold both our struggles and our relative power, for better and worse? This space for our growth and reflection is what I hope to offer in these pages.

PART ONE

Air

To be Black in the imperial core* is to make meanings out of the things we can't see – to read between the lines.

We've learned to find ourselves in the silences between words and histories rarely told. To feel the chilling truths behind *business as usual* and assess rooms by the wordless tension they hold. We understand when a situation just feels . . . off, operating through gut and inkling, a familiarity of ill feeling.

After all, emotions are things you can't touch, and yet paranoia and obsession have been made real enough to kill us – hot rage pulsing through police and state violence, taking our lives over phantom threats that evaporate in the cold light of day. The wrong look or kind of car. Buying sweets. Being at home. Walking down the street. Idling – wrong place, wrong time.

The first part of this book is about the unseen – how we feel about the climate crisis, in all its complications, and an

---

* 'Imperial core' is a term that describes countries, such as in Western Europe, the UK, Canada and the US, who have benefited from colonialism and, as a result of this exploitation, are some of the richest countries in the world.

invitation for those feelings to come to light. We'll also explore air pollution as a perfect metaphor for racism, often invisible to the human eye and easy for many to ignore, but sickening and ultimately deadly.

I.

Feeling Our Way Through the Climate Crisis

1.1: What are we saving?

You wake up. Is there time for breakfast? Coffee? The gym?

Go to work, log in, small talk, check emails, meetings, cup of tea? Lunch, sometimes while working. Your boss criticises you – was that unfair, or did they have a point? Is their racism showing? Having to wade through the difference is the work you'll never get paid for. Check emails. Clock-watch till the socially acceptable time to log off. Forty hours a week like this, and still not enough money to fill the fridge. How many of those hours are lost drowning in office politics, worrying? Day-dreaming? How much time, travelling there, back and forth? What else do you have time for? Pleasure? Friends? Family? Evenings are short.

The pandemic continues, and people are still dying, though no one talks about it. Instead, we say it's over.

Genocides are happening in the Democratic Republic of the Congo, Sudan, Palestine. Atrocities we don't notice, and others we do. We protest against people dying, and the governments – who back these wars – call us the terrorists. At this point, there is no evidence they value any human life other than their own.

They speak for us and our taxes. There is a two-month wait for a doctor's appointment.

None of this life belongs to you, none of this life is yours.

Let me explain.

If you feel like much of your life is going through the motions – giving pieces of yourself to the responsibilities of your life – if you're struggling, like many of us, to keep your head above water, then climate change may well be the last thing on your mind.

Historically, when the climate movement has raised the alarm on the present and impending threats of climate change, it has forgotten those of us already living in a state of crisis. The cost of living, poor health care and illness, systemic racism, ableism, the rise of eugenics, misogyny and transphobia mean many of us live in a state of daily crisis. Asking us to get involved in yet another issue – such as the climate crisis, too often positioned as something distinct and external from us – can sometimes feel like much too heavy a load to carry. This is the tension I've often felt when it comes to Blackness and climate change. Maybe you feel it too. All too often, the cries to save dwindling ecosystems have been made with a passion and care that has rarely been extended to us. All too often, it has felt like a movement disconnected from a broader (Blacker) history.

Narratives that call on us to care and 'save the planet' overlook the complicated relationship many of us have with it. Just what are we being called to save, and who? What would a saved planet look like for a Black collective? In a time fraught with deep societal division, what is this collective goal we're all signing up to when so many of us are being left out of the conversation? When solutions and progress attempt to speak in universal terms on behalf of all of us, climate change conversations already start on rocky ground.

When living in a society that consistently shows how little it values Black lives – that limits our agency and well-being – climate change feels like we've been given yet another problem, with no means of being able to survive or solve it. You only have to look at Black-majority countries across the world; they're among those who've contributed the least to the issue, but who are some of the most affected by climate change. Rarely are these voices truly heard on a global platform when discussing tangible solutions for how to combat the crisis – it's all the burden, with none of the power.

I want to avoid oversimplification – we save ourselves on the daily, because we have to, often with next to little support, and, of course, many of us care about the future of the world we live in. You wouldn't be here if you didn't. But our starting point to this work, to our relationship to the planet, has to begin from a different perspective. It can't take for granted our existing relationship to the planet and to place, and the ways in which it's often contested. To want to 'save the planet', a person must feel they belong to it – which is not always a given when colonial history has told Black people the opposite. And frankly, when has saviourism ever worked well for Black communities?

The white saviour complex is a pattern throughout history in which the righteous white person embarks on some kind of noble mission to save the downtrodden, less capable 'native' – in the spirit of purity and generosity, and nothing else, obviously. The white saviour is really just that good. Never mind that nine times out of ten, they created the mess they claim to be saving people from in the first place. Think international aid sent to 'help' countries who were literally decimated by colonial violence. Those very same countries send over aid, but in doing so, maintain a position of dominance. This isn't salvation; it's just a 'softer' form of colonial power.

We live in a time of separation – from our bodies, each other, as well as nature and the broader world. But the climate is as much to do with biodiversity as it is to do with us. We are the planet; we are the environment and we are the crisis. We are part of the ecosystem in need of change. How we treat ourselves and our communities is how we treat the planet at large. So, in order to save the planet, we need to save ourselves and each other.

Who are we and what do we value? What ways of life are we clinging on to that no longer serve us? How do you need to be saved?

We are the crisis, we live in crisis and we are surviving. I want us to remember this as things get harder.

1.2: Whose crisis, whose timeline?

Content Warning (CW): racially explicit language

How do you feel about climate change? Like, *really* feel?

Beyond the stock answers of climate change being 'really bad' or 'worrying', there's a lot to unpack. Conversations, protests and headlines are showing us that this is a topic of real concern. But the African continent – and many of our countries of origin in it – has long been disproportionately affected by the crisis for decades.

The oil exploitation and subsequent oil spills in Nigeria's Niger Delta by Shell, beginning in the late 1950s, have made this region one of the most polluted in the world.[1] Eastern Africa has a long history of drought, particularly Ethiopia, Somalia, Djibouti and Kenya, which has threatened their food and water security, killing and displacing millions of people.[2]

Maybe you already knew this – if you're engaged in what climate change looks like across the world. But until relatively recently, any reporting of this from the mainstream media would have been reduced to an image of a starving child on dry, cracked ground, looking for food. This has been the formula for a long time – the image of 'helpless African' with jaunty songs like 'Do They Know It's Christmas?' from the 1980s charity supergroup Band Aid thrown in for good measure – but rarely have these depictions been accompanied by explanations of a wider cause and deeper context.

These droughts are a direct result of climate change,[3] which is largely driven by multinational oil, gas and coal companies exploiting the world's resources in the most unethical and harmful ways. Too often, the language around climate crisis suffers from a lack of accountability and has overtones of passivity. It tells us these terrible things are happening, with little rhyme or reason as to why and by whose hand.

Terms like 'crisis', 'ecological breakdown' and 'emergency' are now being more commonly used to hone in on the seriousness of climate change. These words are sharper, more alarming and have paved the way for the emergence of concepts such as 'climate grief' and 'eco-anxiety'. These terms describe a spectrum of emotions, including fear, depression and stress around how the climate crisis is impacting, and will impact, the planet.[4]

I get the idea of eco-anxiety completely. I look at the news and hear yet another grim prediction of how we are living on borrowed time, and I'm terrified at the idea of what this planet will be like in ten or twenty years from now. The heatwaves of 2022 were atrocious, and I worry that they will become much more common in years to come. Sometimes it feels like we're just watching an existential clock ticking away while little is being done to tangibly change the situation. There's the biggest issue: the inadequacies of big power and their lack of action, yes, but what about us? Will we look back and know there's something more we should have done when we had the chance?

I understand the place for climate grief, but it too often feels conditional, especially through a racialised lens. When creating language and new words to describe the human condition, someone inevitably has to notice a set of recurring experiences and give it a name. But who gets to cry urgency and grief, and whose pain and experiences are ignored long before the linguistic flag is raised? Was there not already grief for the Ogoni people of the Niger Delta – who have one of the most significant marine and wetland ecosystems in the world, which is now, thanks to Shell, one of the most oil-polluted regions worldwide?[5] Has there not been ecological breakdown and anxiety for the people of East Africa, living in terror of a four-season drought?[6] For many in the Black diaspora, that anxiety is also felt in the risk of deportation or yet another Black person killed at the hands of the police. That grief is already felt in the vivid images of yet another murder circling around our social media feeds, followed by inaction or a not guilty verdict. I could go on.

In her essay 'Swimmer',[7] writer Nicole Dennis-Benn describes her bewilderment as her white friends debate moving to another country, months after Trump was first elected as President in 2016. Dennis-Benn points out that the terrors of white supremacy her friends feared were *already* her America – a reality she'd had to navigate without the luxury of leaving. She realises that her friends were less empathetic towards the realities of those surviving the racial violence of the US and were, rather, grieving the potential loss of their empire and the freedoms it afforded them. Similarly, the calls for outrage over the shrinking reproductive rights in the US, and ongoing comparisons to *The Handmaid's Tale*, while rightful, ignore how many Black, Brown, Indigenous and disabled communities have had their reproductive rights violently taken away long before now. It is only understood as a standout moment, now at this point in history, because of its capacity to affect a white, non-disabled population.

I write this because it mirrors the issue with climate change and the emerging wave of climate anxiety – it's whiteness grieving the loss of certainty around home and security that so many of us have never had access to. Anxiety over how our planet is changing is real. But the unease feels deeper than this; if it were simply about the loss of ecosystems in the world at large, there would have already been mass outrage and a collective sense of anxiety and grief. Professor and author Sarah Jaquette Ray also reflects on this in her article 'Climate Anxiety Is an Overwhelmingly White Phenomenon', noting that progressives who claim climate change as one of the biggest 'existential threats of our time' blatantly ignore slavery, colonialism, police brutality and other existential threats many of us have had to deal with long before now.[8]

I've come to understand that this is not just the role of whiteness, but of privilege. Many people who experience chronic illness and disability later in life, for example, only become aware of the deeply ableist society we live in – one in which disability is undermined and made invisible – after they are personally affected. The parent raising a trans child only becomes more acutely aware of transphobia when they have to face the realities of navigating society with their child. Living in silos separates us from one another and also divides our struggles. It then becomes a choice to listen, engage and be in solidarity with one another. Most of us don't make that choice until the struggle comes for us.

The rise in conversations around the climate crisis and eco-anxiety is because the waters – metaphorical and literal – are flooding into the white experience. Smoke from wildfires, flooding and unbearable temperatures aren't just issues that can be watched and ignored from the safety of a screen anymore; they are an increasingly lived reality.

For those experiencing life outside of a whiter, safer experience, there is a need to reconceptualise our emotions around climate change. Our timelines for understanding what the crisis means might be very different, overlapping and contradicting the movement at large. For starters, how much have we ever been allowed to truly process, even panic about, the crises in our lives while constantly navigating them?

I feel many things beyond the mainstream narrative of eco-anxiety. When the urgency of the climate movement is handed to me, but so often overlooks those who have long been affected, I feel disinterest and frankly, irritation. I once gave a talk on how Black and Brown lives are affected by climate injustice, and the added violence of these experiences being ignored and unacknowledged in favour of other climate narratives. When it was time for questions from the audience, someone asked me – without irony – to talk about the HS2 high-speed rail network, a proposed government development that has the potential to threaten woodlands and wildlife in its construction.[9] Maybe there was a connection to be made, but in the moment my mind shut down. The repeated othering and downplaying of the concerns of Black majority countries and their diaspora is a barometer of how little the humanity of Black lives *seems* to matter to the mainstream climate change movement. When these histories are missing from the narrative, so much is glossed over – history that society must acknowledge and remedy if we actually want meaningful change.

Many of our family members living in our countries of origin have to deal with colonial violence through resource exploitation and the environmental devastation that happens as a result. The Democratic Republic of the Congo (DRC) has been in a state of conflict and unrest since it came under King Leopold II's control in the late nineteenth century and was later formally colonised by Belgium. Rich in natural resources like oil, gas, diamonds and gold, it has attracted the eyes – and violence – of many colonial powers from decade to decade, who have used every possible tactic to strip the country of its resources, fuelling armed conflict, sexual violence and ongoing genocide in the process.[10]

The environmental impacts of this are also significant: ecosystems have been destroyed, leading to increased pollution and lower air quality. Coltan is a highly sought-after mineral, which is used in many of our electronics. The majority of the world's coltan mines are located in the Kivu region of the DRC and miners – many of whom are children – run a heightened risk of lung cancer and respiratory diseases after working long hours as forced labour under highly exploitative conditions. Much is still unknown about the full extent of the health impacts they're exposed to.[11] Bodies of water are also being polluted by the mineral separation that takes place in them, which produces radioactive substances, making this both an ecological and public health crisis for both wildlife and people.[12]

The Congo Basin rainforest is one of the largest in the world, second largest after the Amazon, and the widespread deforestation happening there, given its scale, threatens the air quality both regionally and globally,[13] offering another striking example of how we are all connected.

Many people have had to leave to go to some of the very same countries that are funding the genocidal oppression of their country. If that weren't enough, when they come to a country like the UK, they'll encounter dehumanising narratives about who they must be as immigrants in this country. Life will be made deliberately more difficult for them, and they'll live under the constant threat of deportation – either 'back to where they came from' or, randomly, if the Tory party had had its way in 2024, to Rwanda.

They'll likely be buying phones and laptops at inflated prices, made from coltan mined from their homeland, to keep in touch with family members who might still be there, suffering. How ignorant, to ask them – and the many others of us who also live within these jarring dichotomies, both relying on and being destroyed by the same oppressive powers – to engage in a Eurocentric version of climate change that excludes us completely. To call us 'hard to reach' groups and assume our disengagement must mean we just don't care about climate change.

I'm angrier about this than I realise.

It's a rage with nowhere to go, so it hardens. It's stunning how casual violence can be. Much like when someone called me a nigger from the window of their car and didn't even stop as they were saying it (perhaps thankfully), they just threw it out like a piece of rubbish on the go.

While climate justice and the struggles of Black majority countries are becoming more prevalent within climate change narratives, there is a scepticism that many (myself included) feel – one that the broader movement must acknowledge. A more intersectional change is welcome, but it's all so very late.

Jennifer Uchendu, founder of SustyVibes, a youth-led organisation committed to making sustainability more practical and relatable in Lagos, Nigeria, talks to me along similar lines:

> We've been talking about loss and climate change for ages, and it's rarely come to fore as a core issue [globally], because developed countries and the West just can't relate to the idea of livelihoods, cultures and a sense of well-being being lost as a result of the climate crisis.

For Jennifer, also a community organiser, and her colleagues at SustyVibes, recognising the emotional toil of the crisis is a crucial part of the work of climate activism. The emotional impact matters, and when recognised, can be turned into action. This is how The Eco-anxiety in Africa Project (TEAP)[14] was founded. A sub-project within SustyVibes, TEAP is a pan-African initiative that centres the concerns and eco-anxieties of African communities.

Africa is one of the most biodiverse continents in the world, but extreme temperatures and increasingly erratic weather patterns will continue to disrupt agriculture and food stability, leading to rising food prices and worsening poverty. As rural areas are some of the most affected, more people are moving to cities, a pattern predicted to increase substantially over the coming decades. However, many of these urban areas lack sufficient infrastructure to handle an increased population and don't have the adequate levels of housing, leading to ongoing levels of urban poverty and heightened vulnerabilities to heatwaves and flooding.[15]

While the more practical consequences of climate-related changes across the continent can be more readily understood, there are fewer narratives that make space for the emotional toll on the communities living there. This is a jarring contrast to the emerging language around eco-grief and anxiety from the imperial core.

The trouble is, for a long time, Black people have been denied their humanity, and so our pain too often sits on the fringes of narratives around climate change – if it's acknowledged at all. For us, any conversation around eco-anxiety has to be more expansive, taking into consideration the overlaps between race, our lived environments, our sense of place and well-being. It needs to be historical and wide-ranging, and most of all, our pain has to be enough.

In an interview with former Brighton-based arts charity ONCA, Shelot Masithi, a South African environmental leader and founder of climate school She4Earth,[16] calls climate change a crisis with many faces, with no clear guidance on which one to look at first. In reflecting on the trauma of climate change, she speaks about growing up with regular water cuts and shortages while seeing major corporations such as Coca-Cola and Nestlé continue to have consistent access to water. On one occasion, during a water shortage, Masithi went to McDonald's, her only option for food and drink that day, and wondered whether she might have starved had she not had the means to pay for a meal.[17] These experiences typify the realities of the crisis – the way that climate change clarifies the pecking order of survival: who will be fine, and who will have to negotiate and sacrifice. This is also echoed by Jennifer, who explains:

> It's clear our definition of the environment goes beyond mountains and trees. It's more about our livelihoods; it's food, sources of income and how . . . everything happening contribute[s] to my ability to thrive – and with that it broadens the conversation for us.

In light of all this, the recent wave of language around eco-anxiety and climate grief feels like too little, too late, to the point where I'm left to wonder what these terms could teach Black people about loss and crisis. They feel like fancy ornaments: beautiful for show when there are guests, but of no real use in everyday life. Similarly, for Shelot, the language of eco-anxiety feels like an understatement that lacks depth in an African context, where psychological support to help communities is lacking.[18] I'm inclined to agree. How can we work through emotions from legacies of colonial and environmental devastation that have never stopped and, if anything, feel like they're getting worse?

Despite my reservations, Jennifer considers eco-anxiety a helpful umbrella term to hold on to. For her, it acknowledges the mental health impact of all of this, even if some redefinition is needed to tailor it to the Black African experience. It gives space for feeling and is necessary precisely because it has been a missing component for so long. When understood as a spectrum of emotions, it makes way for those who may feel a sense of shame and guilt as well as the rage and anger that both Jennifer and Shelot name within their experiences. Allowing our emotions to come through clears the space for something else – possibly something like action.

Jennifer likens the crisis to an ocean: we're all in it, but some of us are sitting in safer boats, with better safety nets and structures, floating on calmer waters. For those of us clinging on in choppier waters, the need to acknowledge the impact on our mental health is essential, which for Jennifer, needs to be contextualised and understood for Africans by Africans.

By extension, the diaspora living in the imperial core also has to reconceptualise these terms, holding the complex histories and contradictions of living between countries and cultures. That in itself creates a whole other level of positionality in terms of our relationship to place(s), and the complexities we may feel towards it. We carry the anxieties of the countries and family members we've left behind and the grief of living in a country that's been violent towards these same countries of origin and, by extension, us. For eco-anxiety and climate grief to mean anything to us, they must stretch over continents and cultural difference, offering a space to grieve the ongoing assault on both our sense of safety and belonging.

1.3: (Black) Solastalgia and slow violence

I've been grappling with the term 'solastalgia' for some time now. It's a term that describes a form of homesickness, experienced when a person's sense of place and environment has been irreparably changed due to environmental factors beyond their control. Coined in 2003 by the philosopher Glenn Albrecht, the word is rooted in the words solace, desolation and nostalgia to express a yearning for a home or place that no longer exists.

Albrecht first came up with this word to describe the experiences of the mining community in New South Wales, Australia, who were suffering from power station pollution and drought in their community (notice whose experiences are worth naming?). While solastalgia could probably be considered a form of eco-anxiety, it also stands apart for me. It's an expression of aftermath – our relationship to a place *after* the flood, *after* the fire. How we are changed and changing, after the damage is done.

The term, while apt enough, gives me mixed feelings. Solastalgia has a more loaded meaning when we consider the Black experience, and frankly, it feels like colonialism under a different name. After all, colonial violence ripped the familiarity of home from underneath the feet of many people within our lineage, destroying their agency, ways of living and knowing the world. That pain, one of witnessing that kind of violence, of your own land becoming foreign to you and repurposed to serve a colonial agenda, is a pain that also deserves its own word. Particularly as it continues to live on through us.

Given this history, as well as our experiences of migrating to countries that are incredibly hostile to our arrival, for many of us, home is already a complicated idea. Home continues to twist and grow more complicated from government to government and through the parameters they consistently redefine for our living. By nature, people of the diaspora live in a kind of transit, caught between different cultures and ideas, and in highly punitive and racialised climates, we survive.

I speak with Land Body Ecologies (LBE), a global network of thinkers, researchers, artists and academics exploring the links between mental health, land rights and ecosystem health, through the lens of solastalgia. Understanding climate change, and more specifically solastalgia, through a Black lens must take on a proactive quality and acknowledge external forces that cause harm and disrupt our environments, which in turn affect our minds and bodies. I put this to Babitha George, Catherine Baxendale and Sylvia Kokunda, members of LBE from India, the UK and Uganda, who reiterate that while solastalgia is a starting point for their research, it's a term that they too are unpicking.

For Sylvia, the term has offered a framework for the loss that she and the Batwa community have endured. One of the oldest Indigenous tribes in East Africa, the Batwa community originates from mountain forests, to which they have deeply significant spiritual and religious ties. In the early 1990s, the Ugandan government, in collaboration with international organisations, declared the area a national park and site of conservation, evicting the Batwa communities living there.[1] So often, particularly within mainstream climate narratives, we're only shown the positive side of conservation; the idea of protecting land and ecosystems at all costs. You could well see something like this included in a piece about positive climate action – the great pledge to preserve wildlife and land. But the hidden assumption within this narrative is that no one (read: of any great importance) was there to begin with.

Sylvia, who's also CEO of the Action for Batwa Empowerment Group, explains that this is a narrative the government has leaned into, reinforced by the additional income from tourists they attract on account of the mountain gorillas in the area. Meanwhile, the Batwa community has received no compensation from the government and are now seen as squatters for living on the fringes of the area or in the slums in towns nearby. Sylvia explains,

> You end up losing your culture, you have different norms and beliefs. When you are evicted, you become detached and lose most of them.

The knock-on effects of this have been devastating: higher rates of alcoholism and a very low percentage of children enrolled in schools. Most schools are too far, and the fees and uniforms are too expensive.[2] Lack of access to medicinal herbs from the forests and the inaccessibility of health care services also contribute to poor health and well-being, leading to low survival rates.[3] The average life expectancy for a member of the Batwa community is twenty-eight.[4] No language can truly describe the multiple levels of horror here, or the depth of damage, but for Sylvia, solastalgia offers a way to begin speaking to the complexities of the situation.

For Babitha, solastalgia captures a slow form of violence, rarely expressed in many cultures: a series of changes happening to your home that in turn impact a person's sense of agency and mental health. When I think of the environment and slow violence in the UK, I think of areas like Brixton, Peckham, Bermondsey and Hackney in South and East London. These were predominantly working-class, Black and Brown areas whose histories have been eroded by gentrification. Rents have been hiked so high that shops and community centres that were once pillars of the community have been replaced by boujie wine and cheese shops and artisanal bakeries. The relentless chain of ongoing property developments – new builds, knocking down council flats and using up green spaces – has created homes too expensive for the people who grew up in these areas to afford.

These areas have been rewritten, much without the input and consent from the people who live there. This, for me, is a form of environmental violence. When I was growing up, going to Brixton would be an event. There were no shops selling Nigerian foods in Rotherham, where I grew up, and while Sheffield had a few shops here and there, my mum would often make the detour to Brixton if we ever came to London to see family friends. It was a bulk shop, buying lots of food, frozen to last. I remember traipsing after her – willing her to take us home – as we went from stall to stall while she bought stock fish, egusi and scotch bonnets. I wouldn't know what to tell my mum now, where to even begin to explain what Brixton has become. And this lack of language is where the trauma lives – it takes all the words.

A lesser-known example is in Cardiff, Wales, which has one of the oldest Black communities in the UK. Its port was named Tiger Bay by Portuguese sailors, allegedly because of the rippling water across the harbour. By the turn of the twentieth century, it was one of the world's biggest docks, with people coming and settling from all over the world.[5] Tiger Bay rapidly became one of the most diverse areas in the country where at least fifty-seven languages were spoken. The first mosque in Wales was built by Somali sailors in the 1890s and stood next to other religious buildings in the area. Different night life cultures also emerged, with cafés and bars showcasing African and Caribbean music, influenced by the incoming sailors.[6]

Threatened by this, and in particular the increasing numbers of Black men in the area, the Cardiff authorities leaned into promiscuous and derogatory stereotyping, using their powers to grind the area into the ground. The council did this by limiting investment into the area until its buildings became too run-down to live in and were ultimately destroyed. This move, which writer Chris Sullivan calls 'social cleansing' and 'cultural vandalism',[7] was particularly aggressive in the 1960s, with the local economy and the docks long in decline, leading to high levels of unemployment.[8] Forty-five streets were destroyed, including houses and historical buildings.[9] Very little of what the area once was remains today.

The plan was ultimately to redevelop the area and 'reclaim' the waterfront.[10] The authorities behind this saw the area 'as a "blank space" ripe for commercial development',[11] and so long-standing communities were displaced to isolated tower blocks on the fringes to make way for private developments.[12]

The idea of any area full of people with their own histories, cultures and communities being viewed as a 'blank space' is chilling but, unfortunately, not surprising. From the countries divvied up into colonies and the Batwa communities evicted from their homes to the people of Brixton, Peckham and

Hackney, and so many others, history shows the most ruthless ways in which people are pushed to the fringes, and are largely ignored when they protest. This struggle is often overlooked because of the idea of something bigger, a more 'noble' cause than the rights of the people already living there –it's seen as no more than collateral damage. The renaming of Tiger Bay to Cardiff Bay in the 1980s, as part of its regeneration mission, is a striking example of this erasure.

These are just a few examples that showcase the extent to which our environments are consistently contested – from our language and culture to the houses we live in, this is a violence that hits on multiple levels. It's hard to talk about. It's not that the language of eco-anxiety is unnecessary; it's that, as Catherine from LBE stresses, it's largely framed by whiteness, and points to the impending fears of an imagined disaster rather than the lived experience of something that has already happened.

A recurring theme that became apparent in speaking with Sylvia, Babitha and Jennifer is how little mental health is acknowledged in the impacts of climate change, particularly for those facing some of the harshest effects of the crisis. Similarly, in my experience of speaking to Black people about climate change, there's no comfortable common language that includes us in the experience, often leading to indifference or surface-level engagement. Maybe, if there's anything to learn from eco-anxiety, it's not to take on its 'newness', but rather to use it, as Jennifer and her colleagues at TEAP are doing, to acknowledge the harm that comes from our environments – both past and present. It's to allow for the range of emotions that something like climate change can bring, as a means, above all, of pointing to where it hurts first.

2.

Air Pollution

2.1: We can't breathe

CW: racial violence, death of Black people

The sentence 'I can't breathe' makes my stomach drop. It makes me think of police officers with dead eyes, of struggle and pleas ignored – a point of no return. These words are painful and are a reminder of so much: the very real, physical suffocation many lives suffer at the hands of police, but also the ways we struggle to catch our breath through the inequalities we face daily.

These have been the final words of many victims of police brutality, words that have, in turn, been emblazoned by the Black Lives Matter (BLM) movement. Protests and slogans now carry these words in tribute. We've built a language to name the cruelties of police brutality and the ways it disproportionately affects us. This is how we can hold powers to account and call for change, however slowly it takes place. Yet, there are many more invisible ways that the state needs to be held accountable for our inability to breathe.

Air pollution forces us to breathe contaminated air, filled with harmful substances in the atmosphere. This can come from a number of places, but most commonly from vehicles on the road, chemicals from factories, as well as dust, pollen and, in more and more parts of the world, forest fires.[1] The idea of air pollution is both abstract and all-encompassing; air is literally all around us, yet it's hard to grasp, or even think about. If police violence feels inescapable and blatant in its cruelty, air pollution is silent, and rarely on our radar by comparison. This is the nature of violence – it operates within both the known and unknown. State violence is as vicious as a knee on the back of George Floyd's neck for nine minutes, as it is a factory spewing out toxic waste in your neighbourhood.

Air pollution is an invisible chokehold, and a more vicious killer, given an estimated seven million people die from it worldwide every year.[2] In the UK, levels of pollution are so far above air quality guidelines, that it's considered unlawful. So much so that the environmental charity ClientEarth has sued and won against the government three times for its ongoing failure to create and uphold any sustainable action to improve the situation.[3] You might expect the usual knock-on effects of air pollution – respiratory issues and asthma – but its damage goes much deeper than this. Other symptoms can include heart disease, stroke and limited foetal development during pregnancy.[4] Emerging research has also linked air pollution to diabetes and neurological conditions.[5]

It strikes me that there's an eerie parallel between air pollution and ongoing racism – both slow and silent, and a deadly backdrop to our lives. They linger in the air, invisible and unspoken, dangerous for how they can kill us quietly, yet so pervasive and woven into the fabric of our lives. When all we are left with are the consequences, who do we blame, and who will believe us?

2.2: Somewhere to live – A brief history of air pollution

Race and class still matter and map closely with pollution, unequal protection, and vulnerability. Today, zip code is still the most potent predictor of an individual's health and well-being

Robert D. Bullard[1]

Racism is always linked to place, determining who gets to exist where and how.

If you take a walk in your local area, you already know who likely lives in the 'nice' spots – on the quieter roads or near a park – and who is more likely to live next to the motorway or on a high street. Black people, people of colour, working-class and migrant communities are more likely to live in these areas, closest to sources of high levels of pollution repeatedly and for extended periods. Air pollution operates like smoke from a fire; for people closest to the combination of heat and smoke, the impacts are severe to fatal. Those further away will still smell the smoke and feel its effects but will be nowhere near as impacted.

Despite how much I've thought about it, it's taken some time to reckon with how much I've been exposed to air pollution. Unless you're steeped in this work, it's not something that regularly comes to mind. It's almost too obvious – a constant hum, like background music. If I'm walking across a busy road or through an industrial site, I'm far more conscious of trying not to get run over or of being alone in a half-empty industrial ground than clocking how polluted it might be. Understanding it takes effort, context and piecing together the clues after the fact. It takes noticing: *Why does my chest feel tight? Where have I been, and what was there? Why do I feel wheezy?* We're often not called to be embodied – to check in for physical sensations – until they become so noticeable that they overpower us. And for a Black woman navigating a racialised and patriarchal society, feelings of discomfort and wrongness in the body are pretty common across a whole range of realities beyond air pollution alone. Add it to the list.

But when given the opportunity to really think about it, I see the signs of air pollution all over my life. My current alarm clock is the growing morning traffic that I can hear from the main road nearby. I grew up near a busy road and spent over five years living on and around Streatham High Road in London, said to be the longest high street in Europe at 1.8 miles[2] and once voted London's most polluted high street.[3]

While I wasn't exactly thrilled to be living on such a busy road – or with the carnage I often found on my doorstep from the night before – it's what I could afford. To my mind, it was outweighed by the positives: a great location, close to supermarkets, shops and good places to eat, and a train station. It was also seconds away from a range of buses that would take me to Brixton and beyond all day and night. We're not primed to understand where we're living through the lens of air pollution risks – or to question it. And even if we do, we have very little room to make different choices.

This set-up is not new. These patterns of inequality were emerging as early as the eighteenth century during the Industrial Revolution in Europe and America. The methods of producing goods were changing from handmade to mass production via machines in factories. And with it came the start of a heavy and long-lasting dependence on coal to power factory machinery, initially for cotton and textiles, and later, for steel, electricity and automobile industries.[4] Following this transformation, a radical shift occurred in how people worked as well as where they lived, with many people moving from the countryside to cities for employment. As a result, cities became overpopulated, and people were forced to live near the factories they worked in, in overcrowded and unsanitary housing conditions. Not only that, but the housing was often located downwind of the factories, meaning people bore the brunt of the smog and air pollution caused by burning coal,[5] while the rich moved out to the suburbs.[6] Only now is research telling us how dangerous those levels of pollution must have been at the time, with toxic air leading to respiratory diseases and many premature deaths.[7]

Our cities and work patterns have undergone many changes since the Industrial Revolution, but those who continue to live closest to these sites of pollution remain largely unchanged. As patterns of migration have evolved, more Black, Brown and working-class communities add to this equation. It's why so often, following our migration, we've built homes and close-knit communities in run-down areas no one else wanted to live in. Power and wealth dictate where and how people get to live and the quality of life afforded to them, while the poorest are forced to live on the edges. This is by design.

'Sacrifice zones' is a term developed by renowned environmental scholar Robert D. Bullard, in conversation with Indigenous leaders at the First National People of Color Environmental Leadership Summit in 1991, a landmark event that took place in Washington DC, USA.[8] The term refers to how the economic gains of building, say, an airport, motorway or factory in an area are considered more valuable than the health and well-being of the people living within them. These are the places where many of us can afford to live. For example, incinerators, sites where waste is burned and where there are higher rates of odour, noise and air pollution, are three times more likely to be built in low-income areas, where a higher proportion of Black people live. Wealthier (read: whiter) areas, on the other hand, are considered more worthy of care and preservation.[9]

This is how racism operates within systems – the consistent decisions, along lines of race and class, to maintain some areas as more pleasant and liveable, while others are left to rot. Black communities in the US know this all too well, with their history of redlining, a practice where the federal government literally decided who should live where by colour-coding different areas according to race and ethnicity. Red areas, those with a heavy African American population, suffered a lack of investment in public services like green spaces, well-resourced schools and health care services, which kept those areas in a consistent state of deprivation. To this day, former redlined areas are some of the hottest and most polluted in the US, largely because of the air pollution from nearby highways, industrial plants and landfills that were built in and around these areas.[10]

Back in the UK, with the forces of gentrification, rent inflation and precarious housing conditions, many of us are now being kicked out of those communities we once made homes in and pushed further still to the edges.

There are clear-cut divides in who is pushed to live where, and then there are grey areas. What strikes me about cities like London, but really many places across the UK, is that there can often be different socioeconomic divides street by street, with larger homes and Victorian mansion houses, turned into flats, alongside council estates or a high-rise just a street over. The increasingly overcrowded and built-up nature of our cities amid rapid property developments is, in some ways, blurring a divide that used to be a lot starker. Much like the climate crisis at large, we're all affected, and polluted air, by nature, is travelling and corrosive – it can't be contained in one area. This is perhaps one of the unintended consequences of gentrification: the wealthier, more upwardly mobile who come to these areas, at the expense of others, inherit the area's environmental stresses. Still, the areas and homes that are largely available to us, for the most part, don't lie. And the decision-making behind town planning – how it is deemed acceptable for us to disproportionately live closer to sites of great environmental harm – reveals a sinister truth of the low value consistently placed on our lives.

2.3: Something in the air – Rosamund and Ella Adoo-Kissi-Debrah

CW: Racism, death of Black people

Ella was in and out of hospital twenty-seven times over two years before she died. Twenty-seven times.

Air pollution was never mentioned.

Until 2021, I lived close to one of the busiest roads in London: the South Circular Road in Lewisham, the same area where nine-year-old Ella Adoo-Kissi-Debrah also lived before her death. If you know anything about air pollution in the UK, it may well be because of the relentless work of Ella's mother, Rosamund Adoo-Kissi-Debrah. When we speak, it's easy to see the former teacher in her; she has a natural authority and a pragmatism that reflects just how much progress has been made with Ella's case. Rosamund is a woman on a mission: a powerful voice on air pollution, founder of The Ella Roberta Foundation, and clean air advocate for the World Health Organization (WHO).

Her daughter, Ella, was a curious and active child, as well as an avid reader and writer. But just before she turned seven, Ella's life changed. She became lethargic and developed a smoker's cough. Over the next twenty-eight months, Ella frequently stopped breathing during the night and suffered from seizures. She was eventually diagnosed with severe asthma and was induced into several comas in an attempt to help her recover.

I am naturally cautious when talking about these details with Rosamund. How to talk about it? I can't begin to imagine what it must have been like to live through, and I'm aware of the pain Black women are often expected to shoulder, which becomes even more complicated in the public eye. When I ask her how she copes with having to talk about Ella's death repeatedly, she speaks about the importance of boundaries and avoiding certain questions. While the media have been largely supportive of Ella's case, the grief that comes with the repeated questioning and talking about it, of course, lasts long after the interviews are over.

In all the attempts to find the cause of Ella's declining health, it was only in 2013, when Ella died aged nine, that an inquest first suggested her death might be linked to 'something in the air'.[1] With so many unknowns, it was through Rosamund's initiative that Ella's medical records and tissue samples were investigated further. Following this, a report was published in 2018 where evidence revealed air pollution as a significant factor in Ella's death, especially given the unlawfully high levels of air pollution in the area that were uncovered as part of the investigation. I ask Rosamund if there is anything she might have done differently had she known air pollution was such a factor. She reflects:

> The only thing I could have done would have been to move . . . when you live near a busy road, there's very little you can do. We did take a different route to school, but you feel like you don't have any control over it. Individuals are blamed quite a lot. I would have been overjoyed had I found out [about air pollution] then. It suits me to think that way now, but were her lungs too damaged to make a difference?

In the Victorian era, doctors used to recommend the ill retreat to the countryside, to rural areas with good air where patients could rest until better health returned.[2] Today, the remnants of this idea remain for those who can afford to take city breaks for fresh country air. But realistically, few of us can pack it up and 'summer' in the country, even if we wanted to. Extended periods of rest and recovery are becoming increasingly unattainable for most of us – in a way that feels increasingly dystopian. Plus, we shouldn't have to leave our communities or the places we work and live in just to be safe.

Children are particularly vulnerable because their bodies and minds are still developing. They're likely to be more physically active, playing games and sports in outdoor spaces and breathing in higher quantities of air – and more pollution.[3] A shocking example of this is in Tower Hamlets, east London, an area known for being densely populated and heavily congested. Children who live there have 10 per cent less lung capacity than their peers elsewhere in the UK.[4] In Birmingham, research shows that children in primary school could die up to half a year early due to the illegal levels of pollution[5] – which is a major trigger for asthma which, in turn, is one of the biggest reasons why children miss school.[6] It's devastating to realise how our children are being failed by a system that has allowed air pollution to spiral out of control.

In December 2020, Ella became the first person in the UK to have air pollution added as a cause of death to her death certificate. While this signals a step change in how air pollution is understood, and therefore legally recognised, there is still much to consider about how this went down. So much of the outcome of this was due to the sheer determination of

Rosamund, fighting through the grief and uncertainty of Ella's death to find answers for seven years – almost as many years as her child was alive. Ella's case may be the first of its kind, but with over 90 per cent of children in the world exposed to toxic air,[7] it begs the question: how many more children have died from air pollution, without their families truly knowing or being able to do anything about it?

2.4: Are you meditating enough?

The 'weathering effect' is a term coined by public health academic, Arline T. Geronimus, to describe the cumulative impact of daily racism on the health of Black people, which makes us more prone to sickness and disease.

It's a fracture within us – splintering more and more from day-to-day racism and microaggressions, little by little, until our overall levels of health and well-being are compromised as a result.

Despite the resilience that Blackness is often associated with – whether through explicit labels like 'strong Black woman' or through the many levels of BS we're expected to shoulder on a daily basis – the pressure of what we live through has to go somewhere.

That hot flush in your bones when something is racist, but you can't 'prove it', the very obviousness of racism, spun into fickle debate and culture wars. Being undermined by the doctor who won't even look you in the eye. The microaggressions at work you have to duck and dive through. Your ideas being taken and credited to someone else. It has to go somewhere.

It whittles away at you, trying to make you weaker, more malleable, less threatening. It is as intended.

We are not an unreserved well of perseverance. A cliff edge, no matter how immovable, will chip and sometimes crumble when faced with generations of relentless wind and storm.

It's still a cliff – beautiful and powerful – but it's impossible to remain unchanged.

We're told Black people are more susceptible to heart conditions, diabetes and high blood pressure. Doctors recommend a healthier diet and exercise but never seem to ask about the wider conditions of our lives. Instead, health is a problem for you and you alone to regulate. Health practitioners tell us to meditate. Religion tells us to pray. The growing wellness industry tells us that the anxieties about the world we're living in can be solved by daily affirmations, manifestation, ice water baths and 5:00 a.m. mornings.

I do some of these things – ritual and connection to the body are important – but it speaks to the idea of our wellness as an individual responsibility, a set of personal choices and behaviours rather than a reflection of the ecosystem we're operating in. Even if you alone are able to 'manifest' your way out of the difficulties of your life, what does it mean if the rest of us are still struggling? If your wellness depends on the sickness and oppression of others, how free then can you truly be?

No one ever made it to contentment and wellness alone, just like no one ever became unhappy or sick alone. There are a whole set of factors and external contexts that weigh into both.

Things that might affect our health:

The condition of the houses we live in – damp and mould from poor-quality housing. How built-up and polluted our local areas are, poor-quality sleep, financial and time constraints that stop us from eating regularly. Heavy shift work, hostile workplaces, single-parent child-rearing and heavy caring responsibilities without support. Low pay, disconnection from a caring community. How safe we feel in the areas we live in, how much we can rest. Fearing the worst when our children leave the house.

Our immune systems are already depleted by chronic stress, which means that any additional illness affects us more acutely because the body has a steeper hill to climb. For example, people suffering from long-term exposure to air pollution, as Black people are more likely to do, are 11 per cent more likely to die from Covid.[1] Research shows some of the highest rates of death from Covid globally have been in heavily polluted areas.[2] In 2020, Black and Brown people were dying of Covid at some of the highest rates. Taxi and bus drivers, operating at the intersection of air pollution and the pandemic – repeatedly driving through some of the most congested and polluted roads, and in close proximity to a high volume of people – were three times more likely to die from Covid than people working in any other sector.[3]

Health is a collective project. Health is not just what we do – it's also how we're allowed to live.

2.5: The trouble with Southall

Southall is one of the most polluted areas in London.[1] It's also been historically known as 'Little India', because of its large South Asian community, who began settling into the area in the 1950s. Over the years, the demographics of the area have evolved with the migration of communities from Afghanistan, Sri Lanka and Somalia,[2] which is reflected in the area's rich political and cultural history. To this day, Southall has one of the lowest proportions of white British residents anywhere in the UK.[3] It's a busy and congested area, located near Heathrow Airport and home to two incinerators, several asphalt plants and high levels of vehicle pollution. It's also caught in the middle of rapid expansion designed to bring new people in and out of the area, even as its current infrastructure is struggling to cater for its current residents.

Angela Fonso, a teaching assistant and campaigner, describes Southall as all 'rack pack and stack' when I ask her about the ongoing changes to the area. She speaks about the juxtaposition of the high rates of new property developments and the expansion of the Southall Crossrail station, happening alongside highly congested roads, closing community centres, and extreme levels of fly-tipping – something she attributes to local residents in overcrowded rental properties having nowhere else to leave their rubbish. While most of us are affected by but less aware of air pollution, Angela, who moved to Southall in 2006, claimed there have been times when she could almost taste the pollution in the air, due to heavy vehicle traffic.

In 2014, things got significantly worse. The property developer, Berkeley Group, bought a former gasworks site[4] to build a new development, said to make way for 4,000 new homes, retail space and improved rail links into central London.[5] The gasworks had been a site for gas manufacturing by burning coal from the nineteenth century up until the 1970s. As is the case with so many gasworks sites, burning coal created a lot of toxic by-products that seeped into the soil, making it highly contaminated.

Despite protests and campaigns from Southall residents who warned of the environmental implications of building on the site, remediation of the soil began in 2017. This is a process that involves cleansing polluted soil to remove the toxins and reverse environmental damage to the land. Environmental concerns aside, when and how these changes take place is telling. Often these kinds of actions are only carried out to serve a business agenda, rather than for the good of the land and surrounding environment in and of itself. From that same year, residents started to experience air that smelled like petrol – so strong that many were unable to open their windows. They also began experiencing tiredness, breathing problems and mental confusion.[6]

In 2018, as the work continued, Angela's health started to deteriorate, forcing her out of her job as a careers adviser at a community college. She attributes the changes in her health to the college's proximity to the gasworks site. Unfortunately, her story is not unique. With so many others in the area experiencing similar issues, fifty residents came together to voice their concerns. As a result, the community group Clean Air for Southall and Hayes (CASH) was set up to campaign against the redevelopment of the gasworks site and investigate its impact on the community. [7]Angela is one of the campaign leads.

Unfortunately, what's happening in Southall is just one example of many. Not only are similar gasworks developments happening in Liverpool, Brighton, Worthing and a growing number of places across the UK, but this is part of a wider pattern of neglect in areas of existing deprivation.

What's shocking is that, despite the grave consequences Southall residents have to live with, nothing in what's happening is recognised as illegal, and very little has changed. Even worse is that an air monitoring report from the former government department Public Health England (PHE),* found it unlikely that the redevelopment would have any connection to air toxicity and long-term illness. Ealing Council has also said that, according to available data from GP surgeries in Southall, there has been no significant change in the percentage of cancer and asthma patients since the redevelopment.[8]

I spoke to Josh Artus, from Centric City Labs, a research organisation that uses neuroscience and geographical data to understand how where we live impacts our health.[9] He explained that any official assessment of how the redevelopment would impact people's health needed to have taken into account the existing realities of the area and its residents.

Southall is a majority Brown and Black working-class community, many of whom will likely do long shift work at night, given that Heathrow Airport is nearby and a major source of employment to area. These factors affect sleep and stress

* Since 2021, Public Health England has been replaced by the UK Health Security Agency (UKHSA) and Office for Health Improvement and Disparities.

levels. This on top of existing high levels of pollution means an additional stressor – like the gasworks site – adds yet another burden to the community, further increasing the risk of chronic health conditions. For Josh, any assessment that ignores this – and assumes that a gasworks redevelopment of this nature in a place like Southall is neutral – is illegitimate. This is how our stories and our needs are made invisible in plain sight.

So, in short – the gasworks development would be harmful in any area, but in a place like Southall, already under so much environmental stress? Catastrophic. And while any newcomers who can afford the flats in this redevelopment should still be concerned, those who are upwardly mobile, white and largely disconnected from the struggles of the area and are unlikely to suffer in the same way. Given that the flats in this redevelopment range from around £400k to £700k,[10] we can assume a fair number of people who live there will fit this profile.

What's also troubling is the official language used to brush all this under the carpet. Look into any official documentation or publicity on this, and you'll find a masterclass in denial. The gasworks site – ironically renamed The Green Quarter – actually looks good. The website shows greenery in and around the site, and the usual buzzwords white people use to describe Black and Brown people are littered throughout. The Green Quarter isn't just any block of flats, you see, but rather a community space that's vibrant! dynamic! and lively![11] They even pull the classic trope of conveniently positioned Black and Brown children playing and laughing – they've truly covered all their bases. We're being led to believe that all is well and going to plan, hand in hand with the community, which, if it wasn't for the residents protesting, it is.

It's hard to reconcile a site that boasts of being one of the most biodiverse redevelopments in the UK[12] – with thirteen acres of parkland[13] – with the daily realities of local residents living around the former gasworks site. Beyond the glossy promos, residents in this same community have been experiencing growing rates of cancer, dizziness and unbearable tar odours – so bad they often have to change their clothes because of how badly they smell from it. These two sides of the story couldn't be further apart. It's disorientating and a reminder of just how many of us have had to learn time and again how official institutions fail us in the most brutal ways, all while they are protected under the guise of bureaucracy.

As is so often the case, for those brave enough to voice the ugly truth, there is punishment for those who deviate from the script. Angela has been called a troublemaker on more than one occasion, and the very legal, democratic ways that CASH has been holding the authorities accountable have been likened to harassment – language that is both loaded and racialised. For Angela:

> Had I been white, I'd have been seen as challenging, but the language would have been more temperate. 'Harassing' is a criminal offence; I've never harassed anyone in my life . . . being a campaigner has made me realise that . . . [racism] is not going anywhere and is deliberately being upheld to suit people and companies with privilege and power. It's been quite an eye-opener.

Air pollution is the perfect storm of neglect because of its invisibility, which means our concerns can be blamed on other things or ignored. It's an abusive dynamic – because where else are we supposed to get guidance on something as unknown as air pollution? It's a public health concern on which good, transparent information should be widely available to everyone. So, it's easy to get lost in the authority of Ealing Council and other official government bodies because there is often nowhere else to turn. Despite myself, I still regularly get caught up in the trap of thinking, *well, they must know best, right?* before bringing myself back to the difficult reality.

Racism thrives on uncertainty. It relies on loopholes and wiggle room, especially when there is money to be made by ignoring our pain. We can point out the obvious, provide evidence, anecdote after anecdote, but unless there is proof that complies in a specific way with an archaic legal system, it's so often dismissed and undermined. It's a steep hill to climb. This is why, when any recognition or change does happen, it's often long after the fact – at the expense of so much of our time, and all too often, our lives. Angela fears the full truth will only come to light ten to fifteen years from now, when the growing rates of cancer clusters and worsening health conditions in the area will force authorities to investigate the links to the former gasworks centre. This is despite the fact that CASH and many others are calling for this work to happen now.

'We are minority in this country,' Angela adds, 'but also invisible, in that our voices are not acted upon. We are denied and our experiences downplayed.'

However Ealing Council wants to explain away the air pollution problem of Southall, they're still neglecting their residents. Regardless of the source, people are reporting toxic air so intense that it's forcing them to leave their homes. As well as cancer, residents are experiencing eye issues, fainting and nausea[14] – this is still a very urgent public health concern. I find it very hard to believe that if residents in affluent parts of Chelsea, west London, for example, were reporting the same issue, there wouldn't be an immediate investigation and action taken. Instead, Ealing Council is relying on second-hand data to essentially tell their residents there isn't a problem – despite the very real symptoms they are experiencing and the direct link between this with the gasworks redevelopment.

When thinking about the climate crisis, many of us assume – despite no real indication – that some wider power will come and do something to solve it. But CASH and Angela's story is a reminder that we can't wait for the official bodies to do the right thing. Whether it be air pollution or the broader impacts of the climate crisis, we're consistently being shown whose lives hold little value and who can, in their imagination, fall by the wayside.

I'm reluctant to place more responsibility on the shoulders of individuals, because this is about the state's failure and inaction, not ours. But the situation does raise questions: how can more of us become more vocal and louder about how we're living in spaces that – by design – are killing us? Much like CASH, how can we form groups that protest and embarrass the government for its neglect as well as develop our own standards and ideas for what we need in the places we live? In the absence of any moral compass that the government seems to live by, how do we form communities that enable us to develop our own?

2.6: No safe level of air pollution

When I speak to Simmone Ahiaku, a campaigner who's been working on air pollution for over eight years, about possible solutions, I am admittedly drowning in its problems. But she points me to small, local actions: buying greenery for our flats and houses to purify the indoor air, using public transport and walking wherever possible, carpooling with neighbours and other parents for the school run and buying as locally as possible. Protecting our existing green spaces is also crucial, not only to preserve our outdoor community spaces and connection to nature, but also because they help improve our air quality.

The fact still remains, as Simmone also asserts, that this is a systemic issue – bigger than any decision you and I can make. A change has to come from the government at an infrastructural level. Road traffic and diesel usage are major contributors to air pollution,[1] but I don't believe that pushing more people to start cycling or hiking up congestion fees for drivers is the way forward. Not without deep structural change that makes walking, cycling and public transport safer, more accessible and more appealing. Otherwise these measures will be ineffective.

Public transport and stations are also still largely inaccessible for many disabled people. Without accessibility built centrally into every station and public mode of transport, what choice do many disabled people have but to use cars?

Another part of our reliance on cars reflects the lack of local goods and services within reach of our local communities. We wouldn't need to rely so heavily on our cars if more of the essentials we needed and wanted were brought closer to us – and if it was generally easier to do life without a car.

This means developing regular bus, train and tram routes that are interconnected nationally, accessible and more affordable, so that driving actually becomes the less feasible alternative. The charity Greenpeace has advocated for a UK-wide rail pass called the Climate Card – a subscription service that would enable people to play a flat monthly fee to use all national railways.[2] This would be a stark contrast to the very expensive and complicated system we have to navigate now. If rolled out across the UK, it could help reduce car journeys and lower air pollution.[3]

But the truth is, there is no such thing as 'slightly less polluted' air – it's either clean or it isn't. When I speak with Josh, he reminds me that we have been taught to accept pollution as an inevitable consequence of society. But there is no safe level of air pollution – someone will always suffer.

For Rosamund, clean air should be a human right that everyone has access to, and the responsibility lies with our decision-makers – from councillors, local councils and the government to corporations, for whom air pollution is a by-product of their business. It's on the government to be transparent and inform us of the levels of air pollution to acknowledge that some of us are suffering more than others and to take the necessary actions to regulate the situation.

That's a standard we should all expect, and while getting there may be a long and complicated journey, that, without question, should be the ultimate destination.

PART TWO

Borders

Much of our lives exist on the fringes of society; from our histories and ideas to the places we often live in. Within the imperial core, we are placed at the sidelines, with the truth of what we know dialled down. Borders are imprinted on our living, and we walk along invisible lines of non-entry.

The topics covered in this section – the prison system, disability and migration – explore the rigid borders that exist within each of these areas, and the structures violently created to position Black bodies at the most marginalised end. The essay themes we'll explore exemplify how polar opposites are created and upheld within mainstream cultures; for there can be no racial superiority of whiteness without the criminalisation of Blackness, no ableist standard of our bodies and minds without the vilification of disability, and no revered 'kingdom' to protect without the creation of borders.

These are made-up, fluid ideas, yet they're enforced through entrenched systems that make their impact and lived reality very real. The climate crisis, sitting atop these issues, also magnifies these realities and clarifies who, by design, is allowed to suffer. My call to dismantle many of these borders comes at a time when it feels like we've never been more defined by them.

Believing in a future without them requires our biggest leap of faith.

3.

Prisons

3.1: Policing each other

The police are far from popular, particularly within many Black communities, and this is a sentiment that's spreading. Shifts in our collective understanding of police violence have led to the popularisation of the term 'ACAB' (All Cops Are Bastards), with widening calls to defund and abolish the police.

But the truth is, policing is a powerful force of control and oppression that plays out well beyond the criminal justice system or any physical prison building. We live within a carceral culture – a system of behaviours that mimics the principles of policing, where we isolate and punish each other when there is harm.

Even our use of phones to record and document each other parallels a Big Brother-style surveillance culture, which greatly benefits the police. On the one hand, it's brought much-needed visibility to acts of police brutality that might have otherwise been brushed under the carpet. But on the other, at the press of a record button, we're watching one another, documenting each other's lives without consent. And when someone's behaviour is considered particularly strange, amusing or clout-worthy, we're posting it online and shaming them.

We police each other all the time.

We embody both roles of the prison warden and prisoner in everyday ways, replicating the shame, fear and disconnection this creates.

Everyday examples[1] include: the expulsion of a 'troublesome' child from school, becoming cold and distant towards a friend who's upset you instead of communicating, 'cancelling' someone online for expressing views you disagree with, or reporting someone to the council or police for behaviour you disapprove of.[2]

These are just a few examples that many of us have been guilty of. How are you punishing others around you? Who is punishing you?

I grew up in a God-fearing, rule-abiding house. Following the rules was also akin to following God, and any deviance from them was considered the ultimate form of transgression. Keeping your head down and staying out of trouble was the key to surviving this world. Trouble was dangerous, far more cruel and unforgiving for a Black family in a white area. It didn't matter that we'd never had any massive interaction with the police. They were living in our heads.

This meant that growing up – and for a long time afterwards – I listened to any authority obediently. For any wrongdoing, no matter the scale, I was sacrificial to the consequences. I was wrong and they – whether parent, teacher, religious leader or God – were right. This was how I came to understand that asking questions could be dangerous.

Respectability politics, which means living by a set of norms and standards set by a more dominant culture,[3] is a form of self-abandonment. It's an attempt to live under the radar of oppression by swallowing its rules whole and embodying them. If you can't beat the racist narrative, become its ambassador. Become a mouthpiece for who is 'in' – and is deserving of protection – and who is 'out' and therefore deserving of punishment. It's a false distinction because, ultimately, no one is saved within an unjust society and its ways of living. It's cannibalistic, and by natural conclusion, will come for everyone eventually. This is because it was never about what you did or didn't do – it was always about who you are, and this can never be changed. Nevertheless, attempting to conform gives the false sense of security that there is a way to outsmart and survive it. But is living at half-full survival – or dying?

The language and methods of policing are deeply familiar to many of us. Official carceral punishment for acts of wrongdoing is, therefore, a natural conclusion – it's only what we've been doing to one another in some shape or form our whole lives. But these behaviours and attitudes feed into and reinforce the very policing system that oppresses Black communities the most.

Abolitionist thinking calls for an undoing of these mindsets and a shift towards building other responses to harm that don't centre the police or carceral thinking. It's emotional and challenging. We deserve justice from the people and systems that mean to hurt us. I'd be lying if I said I didn't want those who have hurt us the most to fall under the same sword – to experience the same forms of punishment that they themselves have created and have put so many of us through.

Educator and grassroots activist Mariame Kaba also speaks about the pleasure of punishment and revenge.[4] There's a real 'gotcha' element to it – which, let's be honest, can be satisfying – especially for those who have always been the underdog. It's also why, as Mariame articulates, alternatives can often 'cause people to viscerally and sometimes violently fight back against any possibility of something different than what exists'.[5]

But the apparent safety that carceral punishment offers, she adds, is an illusion.[6] Beyond the initial relief or vengeance it might offer, how does it fundamentally change the root cause? Does the rare police officer who gets locked away for police brutality stop the many more who continue to get away with those same harms?

What does imprisonment truly end, and how does it truly heal us?

3.2: Prison ecology – Climate change from the 'inside'

CW: death, neglect

Out of sight, out of mind is a dangerous place. It's where all the worst things happen.

Mainstream society would have us understand prisons as a cure for all social ills. The policing system relies on us buying into the moral assumption that the punishment – whatever form it takes – is justified by the crime. Its power and longevity are due, in part, to the fact that the prison system exists largely outside of the collective consciousness of the general population. Beyond what's imagined of prisons within pop culture, unless you've actually been imprisoned, worked there or visited a family member, there's still an air of mystery around what prison is truly like. But this is dangerous because it means pretty much anything can happen with very little accountability.

The prison system is far from exempt from the climate change conversation; it is, in fact, a sinister intersection of state violence, racial injustice and climate injustice. Prisons have to be built somewhere, and much like how the idea of prisons and carceral punishment live out of our collective imagination, so do the physical buildings. This 'empty space' has allowed for a myriad of environmental and human rights abuses to pass right under our noses, and the mainstream climate movement still has work to do in being a lot more vocal about this.

I speak to Panagioti Tsolkas from the Prison Ecology Project, an initiative of the Human Rights Defense Center, based in Florida, USA. Prison ecology looks at where mass incarceration and environmental degradation overlap, highlighting who is the most impacted by the areas in and around the prison system.[1]

Prisons in the USA are often built in rural and poorer areas, on cheap and polluted land due to lax regulation. These stretches of land may have previously been mining sites or landfill – so already sites of damage and contamination. This often means that people are literally confined for years in a prison, where the water is poisonous because of sewage and toxic waste infiltrating the water system.

In California, a state well known for its prison overpopulation, prison employees and legal staff drink bottled water, knowing it's not safe to drink water from the tap.[2] Yet, prison inmates are often not given a choice, leaving them exposed to legionnaires' disease – a rare form of pneumonia caused by bacteria in the water – as well as cancer and gastrointestinal disorders.[3]

The State Correctional Institution (SCI)-Fayette, a maximum-security prison in Pennsylvania, USA, was built on 40 million tonnes of coal ash – a dumping ground for accumulated piles of waste from coal mining. The wind carries the ash, which contains arsenic, lead and mercury into the prison atmosphere.[4]

Many inmates have reported a range of severe health problems, including shortness of breath, cysts, severe throat swelling, stomach pains, blurred vision and tumours in their mouth and nose and on their skin.[5]

Richard Mosely, former SCI inmate and prison activist, experienced this firsthand and recounts, in an article in *The Ecologist*, how every inmate at the prison site was ill during his time there. Mosely's nose became blocked just days after entering the prison, soon leading to breathing difficulties and digestive problems. Yet doctors would say there was nothing wrong with him, even as his health – and that of those around him – rapidly deteriorated.[6]

Speaking with Panagioti about this is clarifying. It's clear he's seen and heard too much from the imprisoned people he works with to be subtle. Prisons – or, in his words, 'human warehouses' – are something out of a dystopian horror film. Those who work with him on the magazine *Prison Legal News* talk about very traumatic experiences while normalising them – things like seeing people die by suicide or medical neglect. Mental illness is common, but few receive medical treatment – as it is near impossible to receive treatment for anything. Massive mould outbreaks are also a regular occurrence, so much so people don't even talk about them unless you ask.

The UK prison system is no better with one of the highest levels of incarceration in Western Europe.[7] Think of how the Tory government underfunded our public services over their fourteen years in power: libraries have been shutting down, the NHS is struggling, school buildings are becoming increasingly dangerous and falling apart,[8] and we have to navigate the crumbling infrastructure of our railways and public transport system. These are the public services we can see and rely on daily – which have clearly suffered under the legacy of cut corners and de-prioritisation. So, we can only imagine how the prison system, being so far from the public eye, has fallen into disrepair.

First off, UK prisons have become massively overcrowded, an uptrend that's been happening for decades, despite the fact that most crime is declining.[9] In 2023, it was recorded that some prisons were at 99–100 per cent operational capacity,[10] and others were holding up to 70 per cent more people than is legal.[11] This has been happening under conditions so bad that senior staff inside the system say prisons have been in the worst condition they've ever experienced.[12]

In reading around this, I see similarities in the language that Panagioti also used: prisons described as 'warehousing people'.[13] This, in conjunction with staff shortages, means there can be no real provision of physical, emotional or mental care taking place here. People are living on top of each other under cramped conditions, with many kept in their cells for up to twenty-two hours a day.[14]

This is a pressure-cooker situation and a crisis in its own right – violence, suicides and assault are all on the rise.[15] It also means the imprisoned population is incredibly vulnerable when it comes to external threats, like infectious disease or the climate crisis. The combination of both internal and external crises can only lead to catastrophe. Research from 2021 showed that imprisoned people were three times more likely to die from Covid – not just because of the obvious increased risk of infection in such crowded conditions, but also due to the underlying health issues many suffer, caused, in large part, by being imprisoned.[16]

The buildings are also a problem. There are still several prisons in England and Wales that were built during the Victorian era, which means extreme hot and cold temperatures in summer and winter.[17] While, repairs and adjustments have naturally been made over time, the infrastructure is still nearly 200 years old, with many features of the buildings, such as the ventilation and sanitation systems, at a sub-par standard[18] and ill-prepared for ongoing environmental changes. Newer prison buildings are often built using the older Victorian buildings as a template. So, while the building materials may have modernised, you'll find many with the same structure and design.[19] What a damming indictment of how little our attitudes and imagination towards crime and punishment have changed.

Across the world, the weather is becoming more of a temperamental and dangerous factor. As extreme weather conditions become a widespread reality, the need to adjust and adapt to our environments is becoming a louder conversation. Part of this includes how and if people who are the worst affected will need to move. But what about the imprisoned population who can't leave their immediate surroundings? What provision of care will be offered, considering how little has been given in the first place? For Panagioti:

> Climate impacts will lead to increased floods, heat and cold extreme temperatures. The last couple of years have shown how lots of prisons are unprepared for extreme temperature shifts and major droughts, and flooding.

People imprisoned in Orleans Parish Prison, New Orleans, for example, were left behind during Hurricane Katrina in 2005 while the rest of the city fled amid food and water shortages.[20] This was twenty years ago, and yet little seems to have changed. When Hurricane Helene hit western North Carolina in 2024, over 550 people imprisoned in Mountain View Correctional

Institution were left abandoned there for five days, with some even reporting water coming into their cells.[21] For five days they had little food, no clean drinking water or light, or any sense that anyone might be coming for them.[22]

The UK heatwaves in the summer of 2022 were a wake-up call concerning what extreme weather can look and feel like, but little thought or provision was given to people who are imprisoned. Across the UK, there were reports of prisons becoming 'sweatboxes', with people suffering from severe dehydration and sickness, and limited water and medications being available.[23] Curtains are banned in many cells, meaning rooms become sun traps, rising to unbearable temperatures. Beds are often very close to lid-free toilets, which quickly creates unsanitary conditions. A man reportedly almost died and was hospitalised after extreme heat led to his throat infection becoming worse and blocking his airway. He went into cardiac arrest.[24]

Rikers Island, a New York City prison site, has already received a steady stream of criticism for having no hurricane evacuation plan for its prisoners.[25] The facility, also built on landfill, is nicknamed 'the oven' for its extremely high temperatures due to the use of steel doors, cinder blocks and concrete, as well as its poor ventilation, which raise indoor temperatures up by 10 degrees higher than the outside. Unsurprisingly, on the hottest of days, prisoners have suffered from fainting, heatstroke and heart conditions.[26] Shockingly, someone even baked to death in a cell that reached over 100 degrees.[27]

In other instances of hot weather, people have had to resort to breaking windows that were sealed shut to create ventilation.[28] Heat is anger-inducing and it increases our heart rate, fatigue, levels of discomfort and irritation – all within the already tense environment of an overcrowded prison.

These stories and experiences will only continue to become more common as the climate crisis worsens. But instead of addressing this, the former Tory government, during their fourteen-year rule, only intensified their 'tough on crime' agenda, expanding the scope of who they criminalised, against all common and moral sense. Protesters, the unhoused, immigrants and refugees have all been swept into narratives of criminality and funelled into carceral environments and prisons under these conditions.

Since 2024, under the Labour government, there has been an acknowledgement of the overcrowded conditions of the UK prison system – but more from a stance of party point-scoring and highlighting Tory mismanagement than addressing the question of why the prison system is so overused in the first place. In his first week as Prime Minister, Keir Starmer tellingly commented, 'We have too many prisoners and not enough prisons.'[29]

Despite Labour's early release scheme, an initiative designed to address the excessively high number of people in prison, it's clear there will be no fundamental change or reflection when it comes to challenging the overuse of the prison system. In fact, neither political party seems willing to address why such large numbers of people need to be incarcerated, even as most crimes are decreasing.[30] Rather than questioning whether the prison system is a moral form of punishment – or how it might evolve as the climate crisis demands – we remain stuck in a rigid system, still rooted in Victorian punishment and thinking.

3.3: Being wrong – A social condition

There's a right and wrong according to a moral truth. And then there is whatever the law is doing.

When it comes to the state, ideas of right and wrong are pliable. They can stretch and bend depending on the circumstance and perceived threat posed by a person. Prison represents the ultimate consequence for wrongdoing, and because, on some level, Black people have always been made 'wrong', there continues to be a close relationship between Blackness, surveillance and criminality.

There was an audio clip that was going around for a time on TikTok, which said: 'When you're Black, you're never really lonely, because there will always be a white person all up in your business.'[1] It's been used to accompany videos of Black people being stared at by 'Karens' in public bathrooms, on the beach, in restaurants and shops. Because to be Black in the imperial core is to be under constant surveillance and suspicion. We are, after all, descendants of the countries and people who have been exploited, and we're the underbelly of the imperial core's ideas of superiority – a reminder of who and what must stay in their place for those hierarchies to be maintained.

I was generally considered a good (read: obedient) child at school. But one day in Year Five, I was singled out repeatedly by a supply teacher. I remember the feeling of panic that my usual 'goodness' offered no protection. I'd been quiet, I was following the rules, and yet I was still being watched, to a microscopic degree by the teacher, in ways that felt unbearable.

I was an easy target, as the only Black child in the class, and she had made a decision that I was 'wrong'– and she was sticking to it. So, according to her, I wasn't moving from one side of the classroom to another quickly enough, and everything I whispered to my neighbour, in an otherwise chatty class, made her snap her neck and glare. I was being disruptive, even though I didn't know how.

The breaking point came when I changed my mind about a painting I was doing and decided to change the background from dark grey to white. A wildcard move, granted, but otherwise harmless, resulting in nothing more than a very wet greyish background that was bleeding into the picture. That's it. I keep on searching my mind for more, because it seems so ridiculous – a grey, watery painting. Somewhere, there's someone with a man bun and chinos making bank off the same idea, I'm sure. But for her it was the last straw – reason enough to give me detention, alone in a classroom, over lunch. I remember hearing everyone else playing outside while I sat there, still, too afraid to be wrong.

I want to say that things are changing, that this is an old story – but the root of this thing remains unchanged.

There's a pipeline to imprisonment. It starts young and is pre-empted by a fictionalisation of behaviours that lay the groundwork for the state to get involved. Detentions, suspensions, expulsions and Pupil Referral Units all litter

this pathway to incarceration. To justify these punishments, Blackness has to be made wrong, and positioned
as divergent from the norm, twisting ideas of rightness
and wrongness. So, you create stereotypes and you stick to
them: we're the natural troublemakers you need to keep
an eye on, we're angry and aggressive, Black children are
coded as threatening adults, and God forbid there be a
group of us – it might as well be a military situation.

That's the thing with fear: it has a snowball effect. You plant a
seed of a lie, create enough hysteria and it will just keep growing.
The truth becomes irrelevant. So, the rules you make, and the
harsher punishments you give become justified, as do the theories and falsehoods of inferiority, the pathologisation of our
behaviours and our mental health. In the end, it's for our own
good, all that crushing discipline. However else will we learn
to yield?

This trend of over-criminalising Black communities continues
as a form of control and should be understood as the enduring
legacy of the slave trade. When slavery was outlawed in 1865 in
the US, those who committed a crime were still seen as 'slaves
to the state', meaning that doing anything considered illegal
revoked any human rights you might have and made you as
good as an enslaved person.

Shortly afterwards, habits that were associated with Black
people at the time – such as walking at night 'without
purpose' or hunting on Sundays – were criminalised in an
attempt to redirect Black people back into imprisonment and forced labour.[2] Notice how leisure and slowing
down were criminalised here. To be useful is to be in service,

and our value under whiteness has only been in our labour. By the 1870s, 95 per cent of imprisoned people in the South were Black.[3]

While the UK conveniently hides behind the US when it comes to its histories of slavery, ideas of racial inferiority, Blackness and criminality continue in Britain too. Black people make up only 4 per cent of the population in England and Wales, and yet 13 per cent of prisoners are Black,[4] despite many long-standing protests against racial bias and institutional racism.

When I speak with Adam Elliott-Cooper, lecturer and author of *Black Resistance to British Policing*, he's also quick to highlight the roots of colonialism in how we understand carceral punishment today. Many of these policies and practices were pioneered and tested in colonial contexts. The development of modern prison systems, for example, took place in Jamaica as a means to capture and punish enslaved people who sought freedom. An enslaved person's rebellion – being a fight for freedom – was either met with imprisonment or death.[5]

Also, aside from feigned ideas of morality, prisons make money. This happens because of the prison-industrial complex, an idea that explains how both government and business create profit, through creating and sustaining a culture of surveillance, imprisonment and policing.[6] Just like slavery, this is profitable. There's money to be made by construction companies building more prisons, by the catering companies that supply prison food, by security and surveillance services, or even by external companies that can exploit the cheap labour of imprisoned people and so on. If you're a business that gets a prison contract, in a sector that's only expanding, you're set.

From prisoners being forced into chain gangs to build roads and clear land,[7] till today, the prison-industrial complex allows both state powers and global corporations to exploit prison labour and pay people well below minimum wage to work for them. This is because corporations know imprisoned people have few rights, with little to no access to employee benefits, such as health care, sick leave or a pension.[8]

For example, around 30 per cent of California's wildland fire-fighters are people who are incarcerated.[9] This is high-risk work, which, given the climate crisis, will only become more frequent and dangerous. The job involves using hand tools to clear surrounding woodland to starve growing wildfires and stop them from spreading. If we can imagine them working in among the LA wildfires of January 2025 especially, it seems inconceivable – particularly given earnings are as little as $2.90–5.12 a day.[10] They're also nine times more likely than other firefighters to suffer from smoke inhalation, burns and dehydration.[11]

Wherever there is an economic advantage to keeping the prison industry going, it ceases to operate from any kind of moral framework, and where there is exploitation, it will always replicate existing patterns of oppression. The growing criminalisation and imprisonment of the unhoused for sleeping in public areas is a worrying trend we should keep on our radar. We're seeing increases of this, particularly in the US.[12]

There are also growing fears that the US government's threats of mass deportation of 'illegal' immigrants wouldn't be a deportation out of the country per se, but yet another way to incarcerate migrants and inject more free labour into the prison labour industry.[13] Emerging information, in light of the US's intensifying 'immigration crackdown', seems to confirm this. Tens of thousands of immigrants are being taken nationwide – largely to the southern part of the country – to privately run Immigration Customs and Enforcement (ICE) detention centres.[14] Chilling.

While the prison-industrial complex has been popularised in the US, it's also expanding its web in the UK. Private security companies such as Serco, Sodexo and G4S run a number of private prisons in the UK, making it one of the most privatised prison systems in Europe.[15] Unsurprisingly, where business is prioritised over social welfare or rehabilitation, corners are cut for the sake of profit margins, which worsens the levels of overcrowding, limited staff numbers and violence that the prison system is already dealing with.

This is a part of public life that the world of business and profit has no right to be a part of. And yet, economic incentives drive the prison system forward, expanding it, especially in ways that intersect with race. Adam Elliott-Cooper explains how this can be seen in the newer categories of crime: terrorism, crossing borders, 'illegal' immigration, drug use and drug sales – crimes and sentences that exploit and manipulate Black and Brown populations, further adding to the prison system as intended.

The legal system is not our moral guide – quite the opposite. Rather, it's a mirror of the corruption, racial bias and exploitation embedded within its structure.

This can also be seen in how gang violence is understood. Joint enterprise is a legal framework under which a number of people who are present when a crime takes place can be convicted under the banner of 'assisting or encouraging'.[16] This alone has the power to consistently reshape narratives around who is right and wrong. With such broad room for interpretation, Black communities too often find themselves on the side of 'wrongness'. For example, terms such as 'gang crime' are already loaded and heavily racialised, but under joint enterprise, more people – however tangentially connected – can potentially be charged with the same crime.

In May 2022, ten young boys in Manchester were found guilty of conspiracy to murder and cause grievous bodily harm based on a number of text messages following the death of their friend, despite there being no murder victim or tangible harm. Within the trial, the teenagers were framed as a gang, making it easier to charge them as a group and to draw on stereotypical violent behaviour.

According to Kids of Colour, a youth organisation that supported and advocated for these boys, some of them were on trial for making and liking drill music, as well as their social media activity.[17]

Compare this with people who get away with white-collar crime or political corruption. The government could 'misspend' £4 billion on faulty PPE [18] during the first years of the pandemic, while countless health care workers died due to lack of proper equipment.[19] Since 2010, austerity measures have killed over 330,000 people, disproportionately harming the disabled and those living within poverty.[20] Yet so often, these deadly decisions are explained away by a task force and a report that leads nowhere.

Police officers get away with literal murder, as well as police brutality and sexual assault. Between October 2021 and April 2022 alone, over 1,500 police officers were accused of sexual assault against women and girls, and yet less than 1 per cent have been sacked.[21] That 1 per cent is so telling. Punishment never runs at the same speed or level of consistency for the powerful. There's often a sacrificial lamb – or in this case, the 1 per cent – who are found guilty, once in a while, to prove the system 'works' but all too often there's no accountability for most of the police officers committing these acts of aggression.

This can't happen in a silo; the wider police system must, to some degree, have been aware of these crimes and yet it protects them. How many people in and around the police force had to have known, yet turned their backs or lied? Isn't this like a gang, protecting itself? Why isn't joint enterprise applied in these instances, given the complicity and culture of enabling that is rife within cases of sexual assault, and that is likely to have taken place here?

We are long past the 'punishment fits the crime' ideology. Both punishment and crime are, in fact, subjective according to levels of power, class, race and ethnicity. It is only pseudo-moral frameworks that position some people as innocent, and worth saving, and others as guilty, and deserving of what they get. But nothing can justify the inhumane prison conditions that many have to reckon with when they get there.

How can we understand our humanity as a non-negotiable right, no matter who it is? I know that's a challenge when we think of some of the worst things a person can do, and there's no doubt, there will always need to be accountability. But if we accept the idea that some people deserve inhumane treatment then ultimately, we open the space for such treatment to be used against us all – and there will never be any justice in that.

3.4: Good trouble – The dilemma of direct action

Direct action is a tool that oppressed people have used to build their power throughout history. When communities don't have billions of dollars to spend, they leverage risk. They put their bodies, freedom and safety on the line

Joshua Khan Russell[1]

The downside to being rule-abiding is that you never get into trouble. Rebellion, particularly for a young adult, is a formative way to come into the world, to understand the gulf between what it demands from you and your own longings. It kick-starts a lifelong journey of carving out the lines of your own boundaries and needs, of understanding your place in the scheme of others, who also have their own wants and needs.

Rebellious highlights of my youth are fairly thin on the ground, but include being dared to smoke a cigarette in exchange for a sausage roll – though really, I was only in it for the sausage roll, and so my takeaway was that I needed more Greggs money rather than a potential nicotine addiction. When I was much younger, I once ran away to play with a friend, crawling under my locked gate, with no way of getting back in. I was also a surprisingly good liar in my time but felt much too guilty to commit to the petty theft my friends were into. My low-level questioning of Christianity, simmering in the background of my upbringing, was probably my most shocking rebellion, despite how silent my dissent eventually became.

Looking back at my relationship with trouble, it all feels very PG. Despite how I often felt, I wasn't nearly trouble-some enough – more often ghostly and overly polite. But learning the power of your own rebellion is entirely necessary. While from the perspective of a parent, caregiver or teacher, there's an ease in raising children who fall in line, we must all be taught just as readily the necessity and power of knowing when to say no, even to the people who raised us.

Many of us have been denied this right because we know just how violently exaggerated reactions to our breaking the rules can be. We're given very few chances to make mistakes, and very little space for redemption. I still think of Pupil Q, a 15-year-old Black girl in East London who was strip-searched by two policewomen while on her period, with no other safe adult present, all because teachers had suspected she smelled of weed.[2]

While there has been rightful mass outrage in response to this racist and appalling misconduct, even if she did have weed on her (she didn't) this was a gross overreaction, highlighting the lack of care and qualification of those teachers. This is a situation, among many others, that reflects the ways in which trouble finds us, often when we are doing next to nothing.

The state and all its oppressive powers want us to be submissive and toe the line. But the times we're living in call on us to be on the side of good trouble. The kind that pushes change towards better material conditions for all of us. As times become more fraught and oppressive, good trouble calls on all of us to be bolder, and to live by a moral truth. It empowers us to distance ourselves from the so-called norms established by people in power, who operate out of self-interest, not a public, collective one. It can be daunting, but we only need remember the lineages of protest and resistance from generations before us. Their protests are why we are here today and serve as a reminder that we also have the same opportunity to pay it forward and draw our line in the sand over the conditions we're unwilling to put up with.

Direct action, and more involved forms of protest, such as sit-ins, disruptions and strikes, by design may break – or at least bend – the law. These actions are loud and disorderly and for good reason: they are meant to shake order, activate and inspire, and cast a light on the grave injustices that have become normalised. But how can we reconcile adopting some of these behaviours, when many of us are being stopped and searched, even imprisoned, based on nothing but mere suspicion?

This is something that Sanaz Raji, researcher and founder of Unis Resist Border Controls (URBC), is only too familiar with. As an Iranian American, protest and direct action have been the backbone of her upbringing and political education. She recalls growing up in the 1970s against the backdrop of the Iranian Revolution, when millions of people protested against the oppressive Shah's regime,[3] along with the eight-year Iran–Iraq war. She began to be a part of the anti-war movement during the aftermath of 9/11 and the 'war on terror', a time she remembers as being particularly brutal for the Islamophobia it gave rise to.

Sanaz, who in her own words is under-documented, has been living, working and studying in the UK for twenty years. She has experienced first-hand the backlash against protest, particularly while studying in UK institutions where she has voiced criticisms of their institutional racism. Much of this led to her PhD funding being withdrawn, putting her immigration status – which was tied to her academic studies – in jeopardy.

The realisation of just how easily a student's immigration status can be weaponised against them led to the formation of URBC: a collective of students, lecturers and university workers opposed to how surveillance, border control and anti-migrant policies show up within academic institutions.[4] As well as direct action and protest, URBC also runs workshops and talks on how border control policies show up in higher education, as well as directly supporting individuals affected.

Central to any type of direct action is honesty about the risk involved and an awareness that for some the risk will always be greater than for others. If arrest and imprisonment are a probable outcome of an action, then who has easier access to legal and financial resources, as well as family support and community? Who will make headlines and who won't, and how can that be mobilised for the cause?

The power of the US student encampments in protest against the ongoing genocide in Palestine lay in the fact that many protestors were Ivy League students. Through the lens of the elite, it's embarrassing to see those who are likely to benefit from the current system the most highlighting its violence and 'acting out' – making this a delicious form of protest. The unexpected and seemingly contradictory can be powerful.

The writer Leah Lakshmi Piepzna-Samarasinha, in their book *The Future Is Disabled* also reminds us of the ways disability activists have used their expertise and experiences to protest. They argue that disabled people's ingenuity enables disabled communities to protest in ways that 'the powers that be might never see coming'.[5]

They recall how the late activist Carrie Ann Lucas pulled out the power control of her wheelchair during a protest against cuts to Medicaid, so that her body and chair became a blockade as the police struggled to figure out how to turn it back on.[6] This was all while she livestreamed it to thousands on her phone.[7] Another example is the ways in which deaf people were able to communicate at protests in American Sign Language without the police being able to understand.[8] In 1997, shortly after former Prime Minister Tony Blair proposed significant cuts to welfare, disabled activists from the now-defunct Disabled People's Direct Action Network (DAN), used red paint in front of Downing Street to write the words 'Blair's blood'. One protester, Sue Elswood, recalls how the police wanted to make arrests, but were reluctant to get red paint on their uniforms.[9]

Any activist group rooted in solidarity work must be acutely aware that not everyone can or should protest in the same ways – in fact, it's probably more effective if we don't. For someone who's the sole carer for their family, disabled or waiting on their immigration status, the risk of just showing up to the protest might be too great, particularly in a climate of rising anti-immigration sentiment.

There needs to be a safety and understanding within our movements, to avoid replicating the same harms we're protesting. This means that we have to get to know one another, particularly for longer, ongoing forms of protest. It also means recognising who might be more readily targeted by the police and organising with this knowledge in mind before, during and after an action. There can be no assumption that all who show up to the protest can and should do the same thing, but rather we must recognise our strengths and capacity, and move accordingly – meaning those who are white and deemed more valuable to the state, move to the front.

Others can work on a more strategic level, organising the types of actions a group might take. Those for whom an in-person protest is too risky can offer pastoral support to the children and young people of those who are protesting, or organise an online teach-in to discuss the historical context of the issues being protested. This could look like crowdfunding to financially support people directly affected by the issues being protested. Direct action is as much about traditional forms of protest as it is about using the full range of people's abilities to make sure we're cared for and working with all capacities, big and small. This is how protest can be done well.

Sanaz has an extensive collection of stories of how whiteness can show up badly within protest if unchecked, enacting its own kind of harm and creating more risk for the rest of us. Particular lowlights include a white person showing up to sell magazines at a protest against the deportation of Nigerian students in Teeside and Manchester Met University. When asked to stop, they packed up and left without joining the protest. Another one: someone at the same protest, who after showing up late, started loudly sharing the group's plans in front of security, and then took offence when being told to be quiet.

As a racialised migrant organiser, Sanaz also recalls how people often want an 'Oprah moment' where she spills some grand narrative around her trauma. A kind of offering in exchange for their solidarity – which should never have to be earned, least of all from those parading as comrades.

When injustice thrives in silence, the cost of speaking out, particularly as a Black or Brown person, can often be life-shattering. But the work of URBC is testament to what can happen when we protest together. It's now a network that gives a voice to the experiences of international students navigating carceral conditions within higher education. The mistreatment and deportation of students could have otherwise been brushed aside as standalone incidents, happening under the radar. But URBC is a network that enables a more visible and a damning reflection of these repeated incidents, connecting them to its part within the bigger context of border control. While the risks shouldn't be underplayed, there's power in our voices, and greater safety to be found in using them together.

We also need to reconceptualise what protest looks like for those of us living at the margins of society. Black people are so often met with the harshest, most disproportionate responses to our personhood that it defies logic. We ruffle feathers and make waves just by being alive. Our whole presence is a protest. A Black trans woman, for example, living in loving commitment to herself amid horrific transphobia and violence, is a beautiful protest.

There were many times, as the only Black person working within largely white-dominated organisations, that just being there felt like a protest, a disruption of some kind. I would often remind myself of this, when the overwhelm of being the only one at times felt debilitating. And as someone among the minority of people continuing to mask in an ongoing pandemic, mask-wearing right now is my most consistent form of protest.

While more traditional forms of direct action are important, we can't deny the risks they come with for those who are the most marginalised, particularly in the high-level policing cultures we live in.

But protest can also be quiet and reflective, and just as much in the choices we make about how we live our lives; how we love ourselves and each other, as well as the decisions we make to buy or boycott a product or brand. Protest is also in protecting our peace, by withdrawing our energy and attention from the people and systems that mean to harm us. It's in the choices we make for our lives on our own terms, not just because they're social norms. It's also about speaking up where it's easier to be silent.

These are long, ongoing forms of direct action, the impact of which shouldn't be underestimated.

We are the protest.

3.5: Prison abolition and the shaky practice of forgiveness

Abolition is about presence, not absence. It's about building life-affirming institutions

Ruth Wilson Gilmore[1]

There are moments in all our lives when we will need to forgive others, as well as be forgiven. For some this practice may come easy and is a way to feel free of the burden of resentment, while for others forgiveness is a precious commodity to be given sparingly.

When under the guise of 'love and light' and 'forgive and forget', forgiveness can feel pressured – and when expected from someone in a position of power, it can serve to preserve the status quo. The belief that, for example, the enslaved Black person's reward was in heaven if they endured their suffering and remained faithful to Christian ideology still unsettles my spirit to this day. Particularly, within Christian cultures, the idea of forgiving people 'as God has forgiven you'[2] can create a spiritual bypass – where the complexities of a situation can be tone-policed into palatability.

Even beyond religion, Black people are often pushed into something that seems like forgiveness all the time. Partly because the sheer scale of racial violence we've endured is simply too hard to keep track of. But also because we're often called to push these harms to the back of our minds, as things that have happened long ago in the past, until it blurs into something that looks like placid forgiveness. But I prefer to believe it as a rage we can't touch.

We all have our own wounds and differing relationships with forgiveness, but often, how we process the violence done to us is undercut by a carceral logic, in which exclusion and punishment form the foundation of how we relate to one another. This happens between us, which reinforces these same behaviours within the legal system. But the law, and all its oppressive undercurrents, isn't going to truly address many of the wounds we still carry from the harms we've suffered or get to the root of the issue.

The stale apologies that oppressed peoples across the world receive from the state typically come generations later, if at all, lacking basic human empathy and often riddled with dry language designed to evade legal repercussions. This is how we come to learn how little justice is available to us.

In 2009, former US president Barack Obama signed a 'Congressional Resolution of Apology to Native Americans' for the irreparable harm done to the Native American people. But the government never announced or drew any attention to it, nor were any tribal leaders or groups consulted.[3] It also included the following disclaimer: 'Nothing in it authorizes or supports any legal claims against the United States, and the resolution does not settle any claims.'[4] How very touching.

Such limitations within our current system leave us with the following pitiful options: delayed apologies (if any) from the state and people in positions of power, with little accountability or recourse to justice for those they've harmed, and for everyone else, a carceral system that judges the rest of us as either right or wrong.

No situation offers us any real growth and understanding nor gets us any closer to creating a real sense of safety in our communities. Not to mention how the police do very little for us, as a racist, sexist and homophobic institution[5] that's unwilling to fairly police themselves, never mind anyone else. A 2024 independent report judged the Metropolitan Police as inadequate when it comes to 'investigating crime' and 'managing offenders'. At its very best, it was deemed as adequate for 'police powers and public treatment', with nothing under the good or outstanding category.[6] I wish I were making this up.

Prison abolition describes a practice that aims to reduce, if not eliminate, imprisonment, surveillance and policing cultures,[7] as well as create alternatives that promote safety and reduce harm. In the words of writer Brea Baker, this would involve 'assessing and replacing any system that doesn't serve all of us'.[8] The two-fold approach here is essential to prison abolition – it's not just about taking away carceral cultures and institutions, but also replacing them.

In practical terms, this might look like the progressive defunding of the police in favour of other services, such as social and community workers, as well as adopting community-based approaches such as transformative justice. Transformative justice recognises that when violence and harm occur, they never happen in a vacuum; rather, there is always a deeper structural root cause – not just an individual one.

Examples of transformative justice approaches include working with people who have caused harm to help them take
accountability under trained mediation,[9] or building up community networks to prevent violence and intervene when it
takes place.[10] Key to this is ensuring these measures don't replicate the policing cultures we're already in, but rather that
they hold these situations with more nuance and enable actual
de-escalation.

Should the police need to be called for an disagreement between
neighbours or a mediator? If someone is experiencing a mental
health episode in public, why should the police be involved at
all, as opposed to a mental health practitioner in that community, working with that person's regular doctor? Should a young
person getting into trouble have the police called on them, or
would it be better to call upon a youth worker or counsellor,
in addition to a safe community of people who know them?
In this reality, community networks, care and social work are
public services that are valued and invested in. If we felt more
in tune with our communities, and with people working with
us more locally, then perhaps the police wouldn't feel like an
automatic go-to for so many of us.

Among the pushback on prison abolition is the idea that this
would mean an immediate release of all those who've been
imprisoned, particularly those who have done irreparable
harm and acts of violence.

I don't want to downplay the seriousness of this consideration, but it also buys into the logic that all who go to
prison are 'bad' and those out of prison are 'good'. When,
in reality, some of the most harmful people among us have

never seen the inside of a prison cell. With shockingly low conviction rates for sexual assault, for example – at just 2 per cent[11] – it's naïve to keep believing this system is about justice at all, especially given the predatory and abusive behaviours of the police themselves.[12] If it were about some noble vision of justice, corrupt police officers, billionaires, politicians and war criminals would be lining the prison walls. And Trump, with his very public list of convictions, would not have been re-elected as the US president for a second term.

This is a subject that can feel emotionally charged and far from straightforward. Even though many of us know only too well the shortcomings of the criminal justice system, there is still an emotional part to all this that lingers. It's easier, in many ways, to delegate the business of 'right and wrong' to something external to us than to navigate it for ourselves. Also, the dismantling of any system we're so entangled in is inevitably painful. If, even just in our imagination, we do away with prisons, what does it mean for people who've been imprisoned for most of their lives? Or for a sentence finally given to someone who has harmed us after a long legal battle? What was it all for?

Prison abolition doesn't mean we have to bypass feelings of anger and hurt, which are important responses to harm and injustice. Many of us are recovering from harms we'll never receive apologies for, and in a world of denial – with little acknowledgement of when great harm is done – our pain is perhaps the only proof we have. But I think abolition is an invitation not to stop there. How might things evolve if we, instead, worked towards a place of reflection on what real accountability could look like in response to someone who hurts us? What could repair look like and feel like instead of further harm in retaliation? What deeper understanding could be cultivated from all sides?

These are slow questions to be explored, sometimes answered over a lifetime. None of this should come at the expense of our dignity nor any personal boundaries formed as a result – quite the opposite.

This brings us back to forgiveness. At this point of late-stage capitalism and the ways it calls on us to exist, we're all enacting great harms towards each other. Evidently, some more than others – but no one is in the clear. Just by participating in this system – the things we do and buy, off the back of the exploitative working conditions of others, the toxic individualism and ways we treat one another to survive, the struggles in our communities that we ignore for our own precious comforts – we are both being harmed and are harmful. Who among us then, under a carceral logic, can claim to be beyond its punishment, expulsion and isolation? We are going to have to cultivate the practice of forgiving one another much more readily, including the harms we've also done to ourselves.

Of course, there are exceptions. I can't tell you what in your life is forgivable, even if I wanted to. This rings true, particularly for those of us suffering under oppressive violence. Seemingly every day, I watch the most vicious forms of cruelty inflicted on people in the world. We're surrounded by an overwhelming number of ongoing examples of people, groups and systems committing unspeakable harms on others. In those instances, I'm very far away from forgiveness – I'm with my rage. And because there is not yet a tidy alternative to the police, in those times of deep anger and desperation, the law is all I have to reach for. Maybe if there were something else that could hold real accountability and grief, there would be another place to fall – beyond what can feel like an abyss without the legal system, however flawed. But perhaps that longing, if only passing, is where abolition can begin.

In more personal instances, where the harm has felt particularly complicated, I have found focusing on processing the pain to be more beneficial. I have chosen this over forgiveness, particularly where it has felt pressured, and more in service of the other person. Rather, I have found it better to spend my energy on acceptance, and feeling and releasing the anger, so that the bitterness doesn't live on in me. These are stepping stones that may well lead to forgiveness, but they can also replenish me along the way. Sometimes, just working to come to a place that's more neutral is all I have, and sometimes I don't even have that.

Beyond individual acts, harmful environments, like day-to-day life under systemic racism, are instances where the extent of the violence is so nebulous and intricate that there's no way of finding its beginning and end. Forgiveness here can feel irrelevant – who exactly is supposed to be forgiven?

I think it's better to work on not embodying and passing those harms on – to develop a commitment to non-participation in those oppressive norms, and to call ourselves and others in when we do. We can choose not to succumb to the limitations of the carceral cruelty and punishment we live under and set our own parameters – if only, for now, within our minds and to each other. This too is a form of resisting policing. So let's build from that place – in ways that defy the so-called norms of our current system – in favour of ones that work for all of us.

4.
Disability

Ableism:
*A system of assigning value to people's bodies and minds based on
societally constructed ideas of normalcy, productivity, desirability,
intelligence, excellence and fitness. These constructed ideas are deeply
rooted in eugenics, anti-Blackness, misogyny, colonialism, imperialism and
capitalism . . .*
*This systemic oppression leads people and society determining people's
value based on their culture, ages, languages, appearance, religion, birth
or living place, 'health/wellness' and/or their ability to satisfactorily re/
produce, 'excel' and 'behave'.*

You do not have to be disabled to experience ableism

Talila A. Lewis[1]

4.1: Bringing disability to the centre

CW: ableism, eugenics, police brutality

One of the many things we don't speak about as a society is disability.

I remember being at an event years ago with an old friend when we passed a visibly disabled person. She casually commented on how hard it must be to be disabled, something that she could only ever imagine, given she would never be disabled herself. I distinctly remember thinking how odd her certainty was, her separateness. But, given how relatively low my understanding and awareness around disability was at the time, I had no words for the feeling.

I've also been a part of reinforcing this separation, of nodding emphatically that accessibility for all people is a 'good thing' but rarely thinking much deeper about it than that. I've participated in the typically woolly leftist thinking – a general sense that everyone is equal – but given no real further thought into what this actually means. I've witnessed and have been a part of the violence that the non-disabled world enacts by seeing disability rights as a very important issue while offering no real thought or actionable consideration.

But 2020 was a real turning point for me in understanding how the world views disability as we all came to grips with Covid-19. I watched in disbelief as the deaths of disabled people were viewed as acceptable losses. People's lives, namely the elderly, chronically ill and disabled, who, Antony Fauci, former director of the National Institute of Allergy and Infectious Diseases, calmly stated would 'fall by the wayside'[1] as a natural consequence of the pandemic. This normalisation – that vulnerable lives aren't worth saving – is blatant eugenicist thinking, and ideas we're still living by, especially with the pandemic continuing today.

Eugenics is the belief in a superior human race, which – no surprises – has historically been the white, non-disabled, neurotypical man, with bonus points if you're young, athletic and blonde. Baked into this ideal is the conviction that all efforts and protections should be taken for this superior race to reproduce and thrive, while any necessary measures to disregard and kill the rest of us are justified.

Central to eugenicist thinking are ideas of goodness and desirability, shaped by racist and ableist biases and beliefs. Those outside of the ideal are understood as morally inferior; untrustworthy scapegoats for all social ills of society[2] – sound familiar? The consequences of eugenicist behaviours, of exclusion and ultimately death, are made to seem inevitable, such is the way of the 'weak'. But what it also does is get us far too comfortable with the idea of mass death, and that there are those more deserving of it.

Every massacre and genocide in history has been underpinned and justified by eugenicist thinking. The Holocaust, for example, in which over six million Jews,[3] as well as disabled, Black, Romani, gay and transgender people were killed, was informed by the Nazi rhetoric that these groups were less than human, their existence an existential threat to the 'superior' race, and their murders therefore justified. The disabled in particular were viewed as 'unworthy of life'.[4]

The violence of countless massacres and genocides of Black nations across the world, under colonial rule, was underpinned by the 'civilising mission' of the white man – as a misguided justification to shape both land and people in his own image. Given the impossibility, this meant wilful mass death and disablement.

Forced sterilisation is also a tool of eugenics, used throughout history in particular on Black, Brown and Indigenous people as a form of population control.[5] Even the 'one drop' rule, the idea that people with even one drop of Blackness, i.e., someone Black in a person's family lineage, are somehow 'ruined', taps into eugenicist ideas of purity around race. Fears that by 2045 the white demographic will likely become smaller in favour of racialised and mixed-race demographics[6] all conjure fears of uncontrollable impurity in the white supremacist imaginary.

While it's easier to see this as something that happened a long time ago, it still happens, usually in an ongoing, underhand way. Forced sterilisation and birth control for disabled people are still happening in at least thirteen countries in Europe,[7] and to imprisoned people too, with secret sterilisation programmes being revealed within the US prison system.[8]

Also, the Do Not Resuscitate measures put on the files of people with learning disabilities during the second wave of the pandemic are a sleeping scandal.[9] This happened blatantly, where people with learning disabilities were literally told they would not be resuscitated should they fall ill with Covid.[10] This is particularly disturbing given the ongoing high rates of Covid infections and lack of precautionary measures taken to protect our most vulnerable.[11]

Eugenicist ideas don't just disappear, they mutate and evolve, and in times of crisis and emergency, the veil of pretence falls. Fauci would never have so casually said the lives of white non-disabled bodies would fall by the wayside, and if he did, it would have made much more than a temporary splash on the news feed. His casual attitide in saying this reflects a widely accepted feeling. In 2020, the scientist Richard Dawkins even tweeted that 'eugenics would still work in practice, were it not for moral grounds',[12] yet he fails to acknowledge that morals or no, we're already living in a terrifying time of eugenics. If we weren't, he wouldn't have been so freely able to tweet such a shocking sentiment.

If ableism underpins the day-to-day structure of society, pushing us towards an ideal we must aspire to, then eugenics is the crass, more punishing sibling, enacting all manner of violence to achieve the same goal. There's no picking this apart from Blackness because, in the words of activist Dustin Gibson, 'ableism is inherently anti-black'.[13]

I realise I'm late to thinking more explicitly about this. While this, in part, is a result of the things I wasn't curious enough about, it's also a reflection of what I was never taught to question and look for. Sami Schalk, professor and author of *Black Disability Politics*, asserts that our understanding of Black disability has to look very different from mainstream white conversations, which, in her words, 'have been developed with little attention to the types of disability most common in poor and racialised communities'.[14] This is why it so often remains at the fringes of our understanding.

I think this is part of the damage oppression does: it strips our struggles down to the singular, when really every act of violence tells a story of intersectional struggle, specific to context, time and geography. The overarching logic, which lives within colonialism, is that ideas of normalcy and goodness are attached to whiteness and abnormality to blackness, which is clearly an ableist idea. And the means to achieve this is through colonial violence, the heavy arm of eugenics. I speak to Stephanie Davis, scholar-activist and author of *Queer and Trans People of Colour in the UK*, who breaks it down:

> Colonialism has created categories of human and non-human, who are often prescribed to premature or social death. This often includes the disabled, Black, brown and poor. We're not seen as people in our full complexity.

Not only are we categorised as non-human, but we're consistently dehumanised to try and maintain this hierarchy. I think this is also what has led to an underplaying and denying of disability within the Black community, and the ways many of us have tried to conform our way back to acceptance. This is what ableism wants from us: our unwavering submission to this standard. It demands that we be non-disabled, neurotypical and that any illness, suffering or difficulty never be bigger than our capacity to show up and work. We will never meet this standard, but it demands we aspire to, and at least die trying, within the push and pull of its contradictions.

The reality is, disability isn't some far away hypothetical, but a condition many of us will encounter – if not intermittently throughout our lives, then eventually. According to the World Health Organisation (WHO), 16 per cent of the world experiences disability – that's one in six of us.[15] So the reality is that all of us will either live with a disability or be in relationships with other disabled people.

The writer and critic Susan Sontag, in her seminal book *Illness as Metaphor*, writes, 'Illness is the night-side of life, a more onerous citizenship. Everyone who is born holds a dual citizenship, in the kingdom of the well and the kingdom of the sick.'[16] This idea of disability as a part of a dual citizenship everyone acquires is clarifying, and contrary to the more dominant thought that disability is merely the shameful underbelly of an otherwise 'normal' society.

We know all too well how many of our behaviours can be perceived as anti-social and can land us in trouble. Those who have an invisible disability, for example, might not have their needs taken seriously, and may be judged as lazy, reinforcing well-known connotations of Black stereotypes. Whether we're being too loud or, for those who are non-verbal, too quiet, both are too often coded as dangerous. Ryan Gainer, a fifteen-year-old autistic boy in California, was shot by the police while having a mental health crisis. The police had been called to his family's home to de-escalate the tense situation, and instead of actually doing that, they shot him five seconds upon arrival.[17] We've learned in the most violent of ways that deviance from the norm can be dangerous and used against us institutionally.

Black people are also more likely to be diagnosed with mental illness, and while it tracks that we, among other marginalised groups, would suffer more from mental health conditions due to living under systemic racism, what this is made to mean through an institutional lens is another matter. Black people are four times more likely to be sectioned than white people,[18] and we have to wonder how much of that is a necessary course of action, versus how certain behaviours such as anger, defensiveness and upset within the Black body are pathologised and seen as in need of sedation.

There's often very little grace for those of us in real need of support, and even less safety for us to be our full selves. Perhaps this is why disability has historically been such a taboo within Black cultures: because we see the consequences of stepping out of ableism so harshly. Because of necessity, those of us who can pretend push through. But there's always a cost.

In living by these ableist principles – this made-up script of behaviours that determine who has any value – we ultimately participate in our own erasure. Let me explain: othering and side-lining disability means that when we're eventually pushed to our breaking point, in the form of illnesses and disabilities we can no longer hide or deny, we are outcast to that same separate place we have cast others towards.

Moreover, without deeper honesty and understanding of how our bodies and minds may differ from the ableist standard we all suffer under, we confine ourselves to internalised ableism and gaslighting. This is where we feel like we *should* be able to keep up or do certain things, despite the fact that the demands of late-stage capitalism are cruel and unrealistic to sustain.

It's hard, because the punitive work ethic, especially for many Black older generations, has been tied to our survival, and our ability to provide for our families and relatives back in our countries of origin. It's been what has saved us from poverty, but it's also been what has disabled and killed us.

Without deeper self-honesty about this, we deprive ourselves of the opportunity to do things differently and find kinder spaces where we can be heard and given necessary space and accommodations. Without this, we're prone to conflating our changing capacities with some kind of moral failing, enacting the violence and unsafety of society upon ourselves. Is it any wonder then that we can't, in turn, be accepting and loving towards others?

4.2: Black disability and climate change

If you are a marginalised person . . . you are the mostly
likely to get a disability, you are most likely to feel
the impacts of systemic and structural ableism

Imani Barbarin[1]

CW: ableism, medical eugenics, death, fatphobia, climate disaster

Being Black within late-stage capitalism is disabling – not just because working multiple jobs and long shifts denies us our health and well-being, but also because Black bodies are put in the most dangerous places. From being experimented on in labs without pain relief[2] to risky manual labour without proper protective equipment, our bodies are some of the first in line for society's violence.

This is something that will only become more common, particularly if fossil fuel extraction continues at its current pace. As these resources become increasingly difficult to extract, corporations are more likely to push their employees into more dangerous working circumstances, whether through deeper ocean exploration or coal mining, where explosions, fires and collapsed mine shafts can occur.[3] A report by the *Guardian* and the newsroom *Drilled* also showed how fossil fuel workers were dying, or being left seriously ill, from inhaling toxic fumes from oil storage tanks, with the oil industry taking no precautions to prevent this from happening.[4]

We're also surviving the impact of the wars that many of our families have lived through, and mourning those who didn't survive – often from within the very same countries that bankrolled those wars. It's a trip to think about, and is perhaps deeper at the root of our trauma than we give credit for. This trauma and anger live on through us, which in turn impacts our mental health, sense of safety and our world view. We're live-streaming genocides – shocking and graphic. Technology allows closer proximity to such violence, but these things have been happening for generations to many of our families. We can't underestimate what seeing these crimes play out in real time does to us – both as current injustices playing out and in the historical triggers they activate.

It's the psychological warfare of racism, always operating in the background, still somehow catching you by surprise. The victims of the Windrush scandal who, decades after living in the UK, were reclassified as 'illegal' immigrants at risk of deportation have unsurprisingly suffered a great deal, with many experiencing mental health problems, attempted suicide, hypertension, stroke and death.[5]

Disability should also lie centrally within our understandings of the climate crisis, and its invisibility of the topic mirrors how the non-disabled world continues to deny and underplay disability. But climate change is disabling, and it's a climate and disability justice issue.

Straight off, disability will become more likely as a result of surviving ongoing climate disasters like flooding, air pollution or heatwaves. Given that we're more likely to already be suffering from illnesses, life stresses and vulnerabilities that come with being a marginalised person, another crisis would be far more impactful for us than those with fewer stressors to think about.

Those of us who experience oppression due to factors such as race, class, migration, disability and sexual orientation will be more likely to live in housing that's unable to withstand and protect us from changeable weather conditions. Poor-quality housing also makes our immediate environments uncomfortable and unsafe, such as poor insulation making houses freezing in cold weather. Damp and condensation can lead to mould, which over time can cause asthma, allergies, respiratory infections and even death.[6]

May we never forget Grenfell, a preventable crime in which at least seventy-two people were killed in a fire that took place in a tower block in North Kensington, west London. The use of cheap, flammable cladding on the exterior of the building was a massive safety hazard, which was later revealed to have been used on over a hundred other high-rise buildings,[7] highlighting just how normalised neglect and sketchy standards of maintenance are for social housing.[8]

Disabled people are often left to fend for themselves when a climate crisis occurs. Any kind of major disaster, whether it be a storm or wildfire, or the human and environmental devastation caused by war and genocide can be shocking to watch, never mind experience first-hand. Have you ever considered disabled people within this? How they too might evacuate, navigate inaccessible roads, a lack of electricity, and find information on what's happening? For many, the answer will be no. This is the result of societal and cultural erasure, and it is violent.

While the non-disabled world might fear the earthquake, flood or bomb, disabled communities have to reckon with both the disaster itself and the reality that few people are even thinking to look for them. This is also where ableist thinking harms everyone. No one knows who might get injured if a climate-related disaster were to strike. Accessible, well-considered ways of being evacuated stand to benefit every single one of us.

When Hurricane Katrina hit in New Orleans, USA, in 2005, disabled communities were abandoned outright. According to the documentary *The Right to Be Rescued*, nearly half a million disabled people lived in areas directly affected by the hurricane.[9] It should be noted that provisions for rescuing people in general were already limited, given the underfunding we know more likely happens in areas where the poor, Black and Brown communities live. These were the people who were largely overlooked during the hurricane,[10] and on top of this, there was no plan to move disabled or elderly people to safety. Nor were there any evacuation plans for those in hospital or care homes,[11] where there was no electricity, food or drinkable water. Much of the hospital machinery had even stopped working.[12]

One of the many people caught up in this was Emmett Everett, a fat* Black man, who was born in Honduras and

---

* I use the word 'fat' here not to demean but as a descriptor that's been reclaimed as more of a neutral term by the body positive movement. It has also been used in the original source.

grew up in New Orleans. He was funny and well-loved in his community and extended family. At fifty, he experienced a stroke that caused paralysis in his legs, and when the hurricane struck some ten years later, he was in hospital for bowel surgery. He was otherwise in good health.[13] During the hurricane, with all external communications down, medical staff made the following evacuation plan: those who could walk were to be rescued first by plane or boat, and those they deemed the sickest and heaviest were either to be rescued last or killed by lethal injection. Emmett, who was alert and anxious to be rescued, was killed by lethal injection along with two others. 'Katrina' was put as the cause of death on his death certificate.[14]

Of all the horrors I've come across, this one leaves me speechless.

It's so distressing to read how disabled people die unnecessarily when climate disasters occur. However, this only exposes the most explicitly violent end of pre-existing attitudes that society would have us believe: that disabled people don't have a right to be cared for or saved. These are the same attitudes that play out in the day-to-day lives of disabled people, in buildings, workplaces and health care facilities that often remain inaccessible to them.

This looks like uneven roads and streets, and buildings without step-free access, which affect wheelchair users and people with mobility aids. It looks like little to no lift access to public transport stations, which means people either can't travel or have extra-long journeys to the few stations that do have a lift. It's health care professionals who refuse to wear masks to protect the vulnerable and immunocompromised – or really any of us – and events that don't include captioning or sign language for people who are hard of hearing.

These are things that are lifesaving and life affirming. When they are denied, whether by the council, health care provider, workplace or organisation, it might not feel as visceral as being abandoned during a climate emergency, but it still serves to gradually shut a person out of their own life. Exclusion is denial of existence, which makes the mistreatment and abuse that follows all too easy to justify.

The lack of care and consideration of disability when it comes to climate disasters is only likely to worsen during any future catastrophes – catastrophes that few countries are prepared for. Heatwaves, and generally hotter temperatures, for example, could be catastrophic for people with Parkinson's, multiple sclerosis or long Covid as high temperatures can worsen their existing conditions and affect their ability to regulate their body temperature.[15] More consideration for people on certain kinds of anti-depressants is also needed, given how these medications can also cause heat intolerance.[16]

How will those who rely on long-term medication, such as people with diabetes, chronic pain or kidney disease, survive if climate conditions affect how these goods are imported? Or trans people relying on hormone treatments for that matter? How much of the medication that many people rely on could survive under extreme heat or if the power cuts out, if their medication normally needs to be refrigerated?

Many evacuation plans rely on our ability to move as and when necessary – but that in itself is ableist and classist thinking. How will those who depend on extensive medical machinery do this, like people undergoing cancer treatments, dialysis or treatments for head and spinal injuries, which rely on care specific to their location? Not to mention people on low incomes or living in poverty, who can't just pick up and drive or book into a hotel.

During Hurricane Helene in 2024, prices skyrocketed, with companies price gouging fuel and food. There were even reports of shops charging up to $60 for a bottle of water.[17] It was the same with the cost of local hotel rooms and flights, with journeys that normally cost $100 surging to $1000 or more.[18] This is a part of disaster capitalism in action – where catastrophes provide yet another opportunity for capitalism to rear its ugly head and make money off of our lack of safety, and our inability, in moments of chaos, to protest. It's unthinkably cruel and turns what are already expensive and nightmarish situations into an impossibility for so many.

Back in this hemisphere, Europe is currently one of the fastest warming continents on the planet.[19] So beyond emergency responses to one-off disasters, with a beginning and end, what's our ongoing emergency response, and who is – or isn't – included?

Within the lull of any tangible action lies a larger, ongoing issue we need to confront: how we adapt and evolve in a way where we truly take care of one another. This isn't something we can ignore or deflect; it's a looming question in desperate need of our attention. Without centring and including disabled people and their perspectives, allowing them to illuminate the shortcomings and cruelties of the non-disabled world, we're far from having the full breadth of conversations needed about the climate crisis, and further still from truly understanding real ways of achieving climate adaptation.

4.3: The miseducation of Covid – A primer

All art is a kind of confession . . . All artists, if they are to survive, are
forced, at last, to tell the whole story; to vomit the anguish up

James Baldwin[1]

CW: realities and consequences of Covid

Fair warning – there will be no talk of us living in a 'post-Covid'
era here. We're still in this thing, with over seven million dead
and counting at the time of writing.[2] This figure is likely to
be significantly higher,[3] as its doesn't include Covid-related
deaths: those who, post-infection, become more susceptible
to illnesses such as diabetes, cancer, brain damage, heart disease and strokes, that may go on to kill them. Instead, the true
figure, according to *The Economist*, may well be somewhere
between 19.1 and 36 million deaths.[4]

For all the horrors of 2020, a time of a general collective
acknowledgement of Covid, the global death toll by the end
of the year was just under two million.[5] The over five million
more known deaths that have occurred since have happened
under the pretence of Covid being over. It's a glaring contradiction that can't be bridged by simply ignoring it. Seven million
is a colossal number, comparable to how many civilians died in
the First World War,[6] and far more than the number of AIDS-
related deaths at the peak of the HIV / AIDS pandemic in 1995,
where just under a million were said to have died that year
globally.[7]

I come back to Covid here, and at various points in the book, because it's something that we as a collective seem hellbent on ignoring. But such a large and ongoing denial of reality points to a wound. This is a virus that's disproportionately affecting Black people, and is a damning reflection on our lack of care for the sick and disabled. I come back to it so that we might examine it more truthfully. I come back to it to disrupt and complicate overly simplistic 'post-Covid' narratives that would have us believe it was a blip for a few months, never to be thought of again.

In 2024, Dr Ken Cadwell, professor of medicine at the University of Pennsylvania, in an interview with *The New York Times* likened Covid to 'throwing a bomb in the body'[8] because of the severe, multi-organ damage the virus can cause. This is just one of many evidence-based narratives that contradict the mainstream downplaying of repeat Covid infections.

Firstly, given how splintered our collective understanding of Covid seems to be, here is a primer on what Covid is:

Covid is an airborne virus, passed on through the air we breathe. It's been ranked by the WHO as a biosafety level-3 pathogen,[9] which places the virus on par with anthrax, tuberculosis and yellow fever.

Think of how smoke behaves – how it lingers in the air and builds up over time. The same can be said for the virus. If an infected person has been in an enclosed, unventilated space for some time, the virus can remain in the air hours after the person has left the room. This is why guidance to stand two metres apart doesn't really do much as a standalone precaution if you're unmasked in an enclosed space for an extended period. Eventually, the room will be filled with infected air, which everyone will breathe in and out.

This is why air ventilation matters, either through opening the windows to increase the air flow and/or HEPA air filters, which are like fans that can filter the air of harmful particles, such as dust, mould, bacteria and smoke, as well as lessen the impact of airborne viruses.[10] It's also why HEPA filters should be a much larger part of our future. It's helpful for Covid, but also many other airborne viruses, allergens and environmental influences, such as air pollution, which lower the quality of our air.

Covid is clever – deceptive even – and its effects are devastating. First understood as a respiratory infection, mostly affecting the lungs, nose, mouth and throat, research has since shown that Covid also impacts brain health and cognitive abilities, the nervous system as well as gastrointestinal and cardiovascular health.[11] It can affect the body's organs for months and years after initial infection.[12] And this is just what we know so far. Covid is still evolving and is mutating into different strains that act in different ways. So, infections happening now may not just show up as the typical cold-like symptoms many of us have come to recognise, but rather as a stomach flu.[13]

What we do know is that the virus damages our immune systems, making it harder for the body to recover from both Covid reinfections and other illnesses. Far from the myth of 'immunity debt' – the misguided idea that a person needs to be exposed to illnesses to become stronger from them[14] – the immune system is actually a finite resource that we need to do our best to protect, not deliberately weaken. The way colds and flus now feel worse for many people, the increase in strep throat, respiratory syncytial virus (RSV) and even pneumonia, the widespread low energy and fatigue, all suggest our bodies can no longer handle what they used to. Leading virologists have even likened Covid to HIV noting its potential to damage the entire body's ecosystem,[16] particularly after multiple infections.

The high numbers of flu and respiratory illnesses during the winter of 2024–25 have overwhelmed a number of hospitals across England, to the point where they declared the highest alert level of emergency, also known as a critical incident.[15] This is not normal, and certainly wasn't happening in the same way before 2020.

Covid symptoms can also be subtle: a scratch of the throat, a sniffle or even no symptoms at all. Around 60 per cent of infections are actually asymptomatic,[17] meaning you might not know you'd had Covid. But that doesn't mean it isn't harmful to your body, or that you're not infectious. Without precautions or testing, you could unknowingly have Covid and pass it on to someone else, causing them anything from mild to severe illness, or even death. This has created a messy web of neglect, where infections are being passed back and forth with potentially devastating consequences.

Covid, which the WHO has labelled a 'mass disabling event',[18] is both a predication and a current reality. Around two million people in the UK have long Covid,[19] and a staggering 400 million have been affected globally.[20] This condition causes prolonged and evolving Covid symptoms for months, if not years, after infection. It is often life-altering and terrifying, and there is currently no known cure.

In speaking to friends with long Covid, I'm always struck by how navigating it seems like some warped version of snakes and ladders – trapdoors, odd turns, feeling better one day, ill the next – with no linear way of moving through it. It makes existing illnesses and disabilities worse, accelerating their severity, often adding a whole bunch of new symptoms in the mix too.

With over 200 long Covid symptoms,[21] it is hard to diagnose because one person with long Covid could have entirely different symptoms from another. But some common symptoms include chest pain and heart palpitations, poor circulation, brain fog and issues with concentration, postural tachycardia syndrome (PoTS), chronic fatigue, increased anxiety and depression, tinnitus and migraines. A few symptoms on a very long list.

In all the misinformation and downplaying of Covid, perhaps the most sinister is the lie that children won't be affected. Children, who've pretty much been left unprotected throughout this crisis, are already bearing the consequences of this neglect. In busy school settings, they're likely to have had several Covid infections, on top of the germs and viruses young people are prone to catching anyway.

The trouble is, young people are less likely to be able to advocate for themselves amid a condition that is already being downplayed. But the number of children experiencing long Covid is increasing,[22] and despite how energetic young people often are, many are reporting limitations in the kinds of activities they can do now.[23] Symptoms can include heart damage, autoimmune conditions,[24] changes to personality and cognitive abilities,[25] lightheadedness and shortness of breath after minimal activity.[26] This is scandalous, and we are sleepwalking through it. We are likely raising a very chronically ill generation, and we will have to answer to them when they rightfully have questions about how we behaved during this time.

Here is what Covid is not: mild, 'just a cold', greatly reduced by hand sanitiser, a hoax, caused by a vaccine, over.

It's also not the hypocrisy of the Tory government and their complete failure to adequately manage the situation, nor is it the traumas and isolation of lockdown that many people have suffered. These are, of course, important; these are our experiences and the stories that colour that harrowing time. But they are contextual and don't speak to the realities of an airborne virus and how it behaves. Covid hasn't simply disappeared because we are (rightfully) pissed off, grieving and tired of thinking about it. In the backdrop of our denial and complete governmental mismanagement, more harm is being done. Covid is in fact thriving, mutating and becoming as, if not more, dangerous than when it started, as our bodies suffer infection after infection.

If any of this is new to you, then it's intentional – we've been lied to. It's in part precisely because of the stigma of disability that the real impacts haven't been more widely spoken about. But also because responding to Covid appropriately would have required a massive ideological shift in how we value and care for one another. This is at odds with the capitalist ideology that tells us we're supposed to be in this life alone, caring for no one more than is economically advantageous to do so.

The initial government response to Covid was a sharp departure from routine Tory practice and policy: lockdown measures, track and tracing, free tests, housing the unhoused,[27] grants for the self-employed, encouraging working from home, furlough and job protection schemes. To have handled the pandemic correctly, these efforts would have needed to be sustained and we, the people, prioritised.

This would have involved consistent isolation and proper tracking for every infection, having our salaries protected in the process, providing masks and tests for free, proper investment in the NHS and ongoing accessible communication about the nature of the virus. It would have required air filtration on public transport and public buildings, appropriate health and social support for people with long Covid and related illnesses, global collaboration to share best practices, research to understand the virus with outcomes actually implemented into society and developing accessible solutions to minimise harm.

While very little of this has happened for us, those who are in power have full access to the necessary precautions, as and when they wish to use them. In 2022, former US President Biden brought his own air filtration system for a school visit.[28] Up until 2024, The White House had Covid protocols in place, with required testing for all those in close contact with Biden,

Kamala Harris and their partners.[29] Months after this protocol was dropped, Biden was reinfected with Covid.[30] King Charles uses a HEPA filter at Buckingham Palace[31] and attendees at the World Economic Forum, a yearly gathering where many of the political elites convene, still require Covid testing. A positive test bars access into the building.[32] This was still happening at the time of writing in 2024, long after initial claims of it being over.

For the rest of us, Covid has been dealt with in true capitalist fashion: as immorally as possible, with cut corners and no thought of the long-term consequences. The government wanted our labour, and so instead of telling us the truth and acting accordingly, we were put back to work. We were told the virus was nothing more than a cold, that we needed to make up our own minds about an unprecedented global pandemic, and by extension, to think of no one else in the process. The wishy-washy decision-making completely undermined the seriousness of the virus, and the language used was suitably vague to minimise any legal ramifications.

Think about it: if Covid was truly over, and the UK had anything do with it, there'd be a never-ending fanfare, a self-indulgent procession of how the government and country reigned victorious, never to be defeated by the virus, with imperialist overtones left, right and centre. There'd be an endless parade of boasting, interviews with all involved and memorials, a Nobel Peace Prize, even. Life would be measurably better than what we have now, and we would all know that this horrible thing we've come to understand as Covid was over.

The current Covid indifference or denial – the idea that nothing can be done, and we just need to 'live with it', has been engineered. Better for us to believe this and crack on than have real questions and demands of the government. There are lots of proven ways to reduce the harms of the virus through routine testing, wearing high-quality masks (otherwise known as respirators), regular vaccinations for those who are able, and ensuring collective spaces have good air filtration. UV lighting installed from the ceiling has also shown promising results in its capacity to kill airborne viruses and reduce the risk of Covid too.[33]

Any of these options are a step in the right direction, but the best protection is all of them in a combined, layered approach. These are proactive, empowering choices; these are what it truly means to live with the virus, beyond the binary choices we've been told are available – of either permanent lockdown or going 'back to normal'. Not to mention that if more of us took these precautions, Covid would be far less of an issue than it is now. Even better, if governments had invested in proper ongoing research and care around Covid, we might have even eliminated it.

For those of us now disabled by Covid, and who are still maintaining precautions, we're facing abandonment on all levels. It's devastating. From government and state to family and friends, we're now seen as strange, unable to 'just get over it', and the accommodations we ask for, too extreme. We're watching everyone get sicker and sicker, needlessly. We carry the heavy burden that many of us will not make it. People casually talk about having debilitating three-month coughs and pneumonia, or that they still can't smell or taste certain foods.

Sick leave is at an all-time high, and people are 'randomly' dying from heart attacks and having strokes. It's like watching a slow-moving car crash.

The most perplexing thing for me is that health care professionals have also stopped masking, which should be a no-brainer if they truly mean to do no harm. Instead, health care settings are now sites of great risk, where vulnerable people can't go without the risk of getting even sicker because of others who are unmasked. In one hospital in Wales, 70 per cent of patients with a Covid infection caught it in the hospital.[34] This is outrageous. Masking in health care settings should be the bare minimum of what we're doing, and yet, health care professionals are often the first to pathologise us, assessing us as overly anxious, ignoring the science completely.

They too have been misinformed about Covid, with many NHS staff being told Covid spreads largely through contact and droplets, as opposed to the truth of it being largely airborne.[35] This misinformation has informed many NHS practices, despite the plethora of evidence confirming the opposite. Is it any wonder then, that many of us might be anxious, when we can't even rely on our health care professionals and in many instances may know more than them?

Also, anxiety is when feelings of fear are disproportionate to the risk in question. Behaviours such as masking and caring about air ventilation are actually entirely proportionate to the reality of the situation, recommended even. But you know what is part of anxiety? Avoidance.

Instead of real interrogation into what life *with*, not post-Covid, could look like, we remain in this weird purgatory of denying it yet knowing things aren't right. Of being afraid yet unable to admit it, or living in the confusion of truly believing things are normal yet exasperated by the ongoing illnesses we're now experiencing, unaware of the real cause. We deny ourselves the opportunity for growth and receptivity to necessary change, and we dishonour those who have died and who are struggling with the virus in our continued erasure.

This is the understanding I write from.[36]

4.4: Covid – The magnifier

When we follow the virus – really any virus – we follow the fault lines of our culture. Like all pathogens, the novel coronavirus was not a 'great equaliser', as some initially called it, but a magnifier of divisions already present in our world

Stephen Thrasher[1]

CW: ableism, climate disaster

Covid is invisible – you can't see it, touch it or taste it. The way we talk about it is also evasive – for some it was a thing that happened to us years ago, while for others, it's still very much happening and shaping the world we live in. There's an overwhelm of academic research telling us about the risks and horrors of this virus, yet mainstream news outlets barely report on it, bar a passing nod to yet another surge of infections.

Some posters of the early pandemic years, telling us to reduce the spread and stand two metres apart, remain on buses, in doctors' surgeries and on shopfronts, yet are largely ignored. Covid resides in the blank spaces around every headline, in the reporting of increased sick leave in the workforce, cancelled shows and disrupted air travel, and in 'mystery' viruses and coughs. If an alien were to descend on the planet, they might not know how to find Covid, but Covid would certainly know how to find them.

For the journalist, academic and author of *The Viral Under-class* Stephen Thrasher, viruses illuminate and worsen existing inequalities within our society. For all the uncertainties of Covid, what is predictable is who it affects: disabled, racialised, trans and working-class communities as well as migrants and people who are imprisoned. All for whom space, rest and good-quality health care are things in short supply.

Historically, Black people have been disproportionately exposed to viruses through colonial violence. The colossal damage of the slave trade and both its intended and unintended consequences are particularly relevant here. In his book, Thrasher details how the spread of viruses such as smallpox, hepatitis B, measles and influenza was a result of colonisation, mirroring the routes of the Middle Passage and the slave trade. First contracted from Europeans, these viruses and diseases would thrive below deck on slave ships in the unhygienic, putrid conditions enslaved people were forced to live in for months at a time. Many did not survive, and those who were contagious were stationed on land, which also spread the viruses and diseases there.[2]

For many Indigenous tribes crushed by colonial violence, their deaths were far more the result of the diseases the Europeans brought over than the physical violence. Imperial conquest was far easier to achieve given the high death rate from infection and the Indigenous tribe's limited ability to tend to and protect the land due to illness.[3] This is how we can begin to build a picture of how viruses and disease have been weaponised as an extension of colonial violence and genocide in what many call biological warfare. Beyond bombs and guns, the eviller among us know to let life's circumstances take care of the rest, engineering those circumstances to be so squalid and unsanitary that the means justify the very same outcome.

Today, through Covid, we can see how those same attitudes play out along lines of race, age, class and disability with troubling clarity. It was Mark Twain who said, 'History doesn't repeat itself, but it often rhymes',[4] meaning that events in history have a relationship to one another, overlapping in their meanings, patterns and impact.

Covid is killing and disabling us at a genocidal rate. Just like many of us are outraged and protesting a number of genocides happening across the world, we need to maintain that same energy in taking the necessary Covid precautions to avoid becoming part of – and perpetuating – the problem.

The general attitude at the beginning and perceived end of the pandemic – that 'only' the elderly and the disabled are at risk – has been used as a nonsensical justification for taking no precaution. This is a terrifying sentiment to normalise because the unspoken part of this attitude is that these are the people we should be prepared to sacrifice.

The vaccine hoarding by European powers meant that millions of Covid vaccines went to waste while poorer countries went without access.[5] These health inequalities are playing out right now, with nearly 50 per cent of those infected by Covid across the African continent now suffering from long Covid.[6]

When the former Tory prime minister Boris Johnson said he'd rather let the 'bodies pile high' than have another lockdown,[7] he embodied the deathly mindset of the colonialist – one that sees death and destruction as a worthy price to pay for their ultimate goal: their own wealth and power. It was Black bodies piling high[8] – particularly at the earlier phases of the pandemic – and it was us bearing the consequences of these choices. Tellingly, when it became more apparent that Black and Brown communities were most affected by Covid, white people became less cautious and worried about the virus[9] – how sinister.

Biological warfare in today's age of shifting yet present colonial powers can look like passivity – policy decisions that, by omission, ignore entire demographics of people. It can also be proactive. There's a reason why Israel denied Palestinians the right to access the Covid vaccine.[10] Neglect kills and is as disabling as physical violence.

For many, a refusal to see the full extent of Covid and its consequences comes from a disbelief that the government could be this irresponsible and negligent. But looking back into our collective history, listening to our personal family histories or even just observing the current treatment of the most marginalised among us, it's very much possible. The government *would* do this, because they have been doing this throughout history, and are enacting all manner of harms, time and again. Many of our social ills actually stem from the government's blatant lack of concern for our safety and well-being.

Understanding that the bottom line of governmental care falls far below what many of us could have imagined is terrifying – but it's also from this place that we organise. Firstly, we must divorce our behaviours from those of the government, moving towards an understanding of what is right and wrong in line with our values – within ourselves and our communities. Whether passively or proactively, in following governmental guidelines around Covid, we personify their neglect and we also live by this bottom line by which cruel eugenicist behaviours operate, where vulnerable and disabled lives are considered a necessary loss in the process of getting 'back to normal'.

During my conversation with scholar-activist Stephanie Davis, we also spoke about how the absence of any strong, unified messaging from the left has enabled a vacuum to form around this, encouraging the mainstream to move towards the right. She added:

> The right wing went out in the streets in lockdown, were loud about being anti-vax and their right to a lack of protection towards each other. It's made the discourse around Covid really prickly and the left haven't responded in a coherent way.

It's true that the left, as a movement, has been largely silent and ineffective about this, and has offered no real difference to the right in its solidarity and understanding of Covid and disability. Our actions must be more values-led, more courageous and distinct than simply following the loudest voices on the right, or the silence and omission of care from the left. These two approaches reinforce one another and are two harmful sides of the same coin.

Covid, in many ways, is a telling mirror – if we could look at it more squarely, it could teach us a great deal, particularly around the climate crisis. Changing habitats due to rising temperatures and extreme weather conditions such as heatwaves, heavy rainfall and flooding means increased migration of people, but also animals, insects and rodents – all potential carriers of viruses and disease. Different people, species and ecosystems shifting and converging together will lead to new interactions and opportunities for pathogens to spread. Future pandemics are highly likely.[12]

Covid mirrors our relationship to crisis and what we're likely to do about it. What does it say about our readiness for a catastrophe, our solidarity and our staying power? How committed are we to acting in ways that are necessary for the protection of ourselves and our communities, even when it's uncomfortable? How attuned are we to see past government messaging and instead build critical resources and information exchanges that inform and empower us? How would we even know to recognise future pandemics, given our readiness to move past the current one?

These are questions we can and should be asking of ourselves when it comes to the climate crisis. And it comes as a terrifying realisation that through Covid we may already know many of the answers, because who we are right now during this pandemic, is likely who we'll be in future pandemics. If that doesn't sit right with you, the option to change your role within this is available to you at any time.

When Covid first entered our consciousness in 2020, amid the horror of what was happening, something promising emerged: solidarity and mutual aid, and a basic understanding of different people's vulnerabilities and needs. It was also a time of increased accessibility with more of us working from home and developing innovative ways to connect online – something many disability advocates had long been calling for.

Of course, it was far from a wholly socially or politically idyllic time. There have always been people cynical and unbelieving of the pandemic. The government's flip-flopping in their decision-making completely undermined the seriousness of the virus. And yet, the shift from where we were in 2020 to where we are now feels like a violent swerve in a completely different direction.

So, what is the truth? Where does our care and commitment to one another truly lie? Who and what do our behaviours align with? Awful things are happening – and more are likely coming. But if we try to pretend or deny this, it's our despair, our ignorance and our harm that will magnify into something far worse. It's by growing our capacity to recognise and sit with what is difficult that we open up the potential for our care, action and transformation to be magnified instead.

I'll be real: I don't particularly love masking – it can be uncomfortable and often awkward to be one of the few doing so in any given situation. But I do it because it's necessary, as much for my own health and well-being as for those around me. I do it to be a safe person for vulnerable people in my community and beyond, and to show solidarity, in my own small way, with the disabled and immunocompromised who have all but been abandoned by society and cast out to stay at home and shield for five years and counting – those for whom an infection is a sure risk of death or further disablement.

More than anything, as the pandemic silently continues, and it gets harder to keep talking about it, I've come to understand masking as an act of love, a way in which I practise non-consent to participating in the ongoing mass disablement and death, worsening amid our denial. It's disrupting the lie. While it can often be difficult, it's also pretty simple. This is a crisis with an answer – that's rare! There are actual clear-cut solutions here, and tangible ways we can reduce harm. While there needs to be far more collective action for things to improve and our lives to be safer, individual choices also matter. It's personal power, and that, so often, can feel very hard to come by.

If we continue to let this happen to us, then we will continue to suffer its consequences and the eugenicist legacies it enforces. But Covid gives us the chance to build something new, and a means to safeguard ourselves for the future. There is no humane reason why we should be caught without the necessary infrastructure and practices of care for future pandemics, particularly with worrying levels of bird flu on the rise.

Establishing proper medical and public health responses to future pandemics as well as being on the receiving end of state decisions made as a result of these crises will be hard enough. While we should pressure our governments to respond to these public health threats accordingly, we're ultimately affected by how they choose to respond. But the social and cultural mechanisms of communication and care during a crisis, on a community level, are practices Covid should have already taught us. Because we protect each other.

If we let it, Covid could be a great teacher, furthering our understanding of how we can better care and show up for one another in the here and now, and beyond.

4.5: Saying what you need

We're living through an age where time is moving so quickly, it's hard to keep up. The structures that have upheld the ways of living we've all come to know are crumbling, and it's not pretty. I was talking with a friend recently about how interesting a time this is – the downside being that we have to live through it and feel all its growing pains. Somehow, something that happened only a few years back feels like a lifetime ago. That weird ritual of clapping for the NHS during the early days of the pandemic? A whole other country. Global political rifts and scandals are happening left, right and centre at incredible speeds, as are global acts of protest. Even the race riots of 2024 began to feel like something in the distant past just weeks and months after, even if the racial tension still simmers barely beneath the surface.

When things are moving so quickly, and at times traumatically, it's unrealistic to think our bodies and minds will not be impacted, particularly when many of these global changes are affecting us directly. The current norms of society would have us believe that business is as usual, giving us very little room to acknowledge and feel the threats of the world, but this is a lie that only suits the establishment. I think we need to allow for blurriness and grey areas, as the world – and by extension we – are in metamorphosis.

The climate crisis is just one of the many major existential events we're experiencing in our lifetime, and is also an event that blurs distinctions – what safety is there to be found on a changing planet? Inequalities aside, we will all have to grapple with how the planet and our climate are changing. What does wellness and health mean for us in all this uncertainty? Covid also blurs the line between the well and the unwell, as the virus is making disability more fluid and interchangeable than perhaps we've ever known before. Our realisation that the systems that are supposed to protect us, in fact, don't must also continue to complicate our understanding around our health and how we need to care for ourselves and each other.

Along with our evolution comes the need for more nuanced ways of understanding the nature of disability. These are times that call on us to witness and accept the realities of our bodies and minds and how they are changing, and to be honest about our needs. The medical model locates disability as an individual issue and, more often than not, seeks a 'cure', whereas the social model of disability understands society itself as disabling, based on people's biased attitudes towards disability and the lack of accessible accommodations. A well-known example often given to demonstrate this is the idea that a person who uses a wheelchair isn't prevented from entering a building because of their disability, but rather because no step-free access has been provided.

As much as the social model has done to challenge our collective understanding of disability, academic and author of *Feminist, Queer, Crip*, Alison Kafer, argues that this model still operates around fixed identities of who is and isn't disabled.[1] Kafer, rather, offers a relational/political model of disability as a more fitting lens through which we might unsettle and add further complexity to our ideas about the subject. A model that, Alison writes, 'sees disability as a site of questions, rather than firm definitions'[2] – fertile for a 'place of activism and collective reimagining'.[3]

I think a political/relational model offers many expansive possibilities that allow ideas of disability to shift and bend across contexts, reconceptualising how we see ourselves and each other. This lens can also span beyond needing an official medical diagnosis, which many of us might not be comfortable with, given how Black people are often treated and pathologised in medical institutions. It may also offer more space for people who are unwell, but who don't see themselves as disabled, as well as those who, as Sami Schalk articulates, have 'been made ill or sick from white supremacist and heteropatriarchal violence and neglect'.[4]

This is a sentiment the writer and author of *How to Tell When We Will Die*, Johanna Hedva, echoes in her essay 'Sick Girl Theory', in which she asserts that '[t]he body and mind are sensitive and reactive to regimes of oppression . . . all of our bodies and minds carry the historical trauma of this . . . it is the world itself that is making and keeping us sick'.[5]

This is not to dilute or co-opt the experiences of people for whom disability is a fixed reality – those who are chronically ill, who rely on round-the-clock care and who survive on long-term social support for their disability or illness. Rather, it's to widen our understanding, dispelling myths of disability as an individual issue and instead acknowledging it as a larger shared reality. The rise of climate disasters is bringing new challenges to us as a society, which is why a political lens is particularly helpful. It's not only that we will be disproportionately sicker, based on how our environments change, but that the political

decisions made will also have the potential to disproportionately disable us.

Under the punitive demands of late-stage capitalism, we're
only as useful as our bodies are in service to the economy.
Those of us with disabilities are erased from society's care,
understanding and attention because we're never supposed to
be wanting, and always giving at the expense of our needs.
Service has been made central to Black identity, especially
Black womanhood. We're supposed to serve, and to the hack-
neyed ideas of the racist, that's all we're good for. Needing and
relying on each other – or even the state – can feel dangerous.
But who are we without our productivity and our usefulness?
Do we have the courage to find the words for how we can be
helped and supported, which is its own form of intimacy, and
way of being known? And are we willing, with same conviction, to respect and listen to the needs of others?

Disability unearths all that we've been told to hide – our
humanness, our need for accommodation, our need to connect and depend on one another. For design researcher, artist
and author of *What Can a Body Do? We Meet the Built Word* Sara
Hendren, 'disability is needfulness, personal, and political,
social.'[6] She adds:

> . . . we might actually say that needfulness – because it is tem-
> poral and changing and over the lifespan – is a feature . . . of
> human life, and maybe even constitutive of the good life, I've
> come to think.[7]

Our need for liveable, safe climates for our bodies and minds to
be accommodated as they evolve and change with the world,

our need to be enough for who we are and not for our labour –
these are unifying needs that underpin the human experience.
How we tend to those needs depends on us. Because when
the bubble of late-stage capitalism finally bursts – we will need
each other, not as a point of shame, but a fact of life. Perhaps
the quality of our lives will be reflected in how we develop
communities where we can both tend to people's needs and be
tended to. A different kind of wealth.

In the words of Johanna Hedva:

> The most anti-capitalist protest is to care for another and take
> care of yourself. To take seriously each other's vulnerability,
> fragility and precarity, and to support it, honour it, empower
> it; to protect each other and enact and practice community.[8]

So, what do you truly need? Where do you feel vulnerable? Let
our needfulness be the starting point for deeper accommodation, honesty and more loving connection.

4.6: Our disabled futures*

Crip[†] genius is what will keep us all alive and bring us home to the just and survivable future we all need. If we have a chance in hell of getting there

Leah Lakshmi Piepzna-Samarasinha[1]

Society would have us believe disability is a source of shame. This is something we all internalise, to varying degrees, by how we judge ourselves and others for our ability to keep up with an ableist standard. But we must kill this shame at the root, culling those ableist ideas as they re-emerge over and over, working, perhaps over lifetimes, to undo our understandings of who deserves a life full of love and dignity. Because simply put, we all do.

We do everyone a disservice by overlooking the disabled – both as a matter of equity and as a matter of learning and survival. Those who are pushed to the margins of society experience crisis first and are already living and surviving realities that mainstream society is still bracing itself for. Many disabled people were masking and developing mutual aid practices before 2020 and have found ways to navigate and survive a society that, more often than not, makes little to no accommodation for

---

* Inspired by Leah Lakshmi Piepzna-Samarasinha's book *The Future Is Disabled*.
† 'Crip' is a term that's been reclaimed by much of the disability movement to subvert its ableist and derogatory understanding.

the full extent of their needs. These practices are pioneering, innovative and ahead of the curve, and it's a damning reflection on society that, through the lens of ableism, this creativity too often gets overlooked.

Responding appropriately to disability doesn't require sympathy or pandering, but rather a pragmatic approach to what is actually needed, and certainly with none of the ableist ideas of what the world and our needs *should* be. Given the likelihood that we will all experience disability, either directly or through a loved one, making accommodations is a way to future-proof our communities for the good of all of us.

Climate change presents us with a number of crises. The government has already been criticised for its lack of planning and action[2] when it comes to rising UK temperatures and heatwaves, which could cause up to 10,000 deaths a year.[3] Air pollution frequently remains at illegally high levels, with very little of this information communicated to the public,[4] and the government's out-and-out neglect of the pandemic means many of us are getting sicker, with no public health measures or precautions in sight. If we're waiting for the government to warn and provide for us, there's too much evidence to suggest that's unlikely to happen, not least because disabled communities, already surviving in the face of governmental neglect, have been telling us as much.

For Stephanie Davis, this is why we need to centre community action as a means of survival. She asserts: 'We need more people taking the reins and doing what our communities need, as our governments abandon us.'

She goes on to mention how Covid has shown us just how easily the science has been sidelined from the mainstream. We are in great need of more people who are invested in learning about the science behind our public health crises and who are able to interpret it at both a community and grassroots level.

This is starting to happen. The People's CDC, a take on the US Centers for Disease Control and Prevention (CDC), is a health–justice coalition that aims to offer accessible science communication, reports and up-to-date guidelines about Covid, where the state won't.[5]

In a similar vein, the emergence of mask blocs worldwide also speaks to community groups forming to fill the void left by our governments. Mask blocs are self-starting groups that hand out free, high-quality masks, Covid tests and, depending on each group's capacity, air purifiers on loan. In January 2025, during the LA wildfires, Mask Bloc LA and community groups such as Clean Air LA, Clean Air Club and Mask Bloc Seattle distributed thousands of high-quality masks and air purifiers to help protect LA residents from smoke and air pollution caused by the fires. Mask Bloc LA alone distributed over 14,000 masks in the first few days. Clean Air LA supplied nearly 200,000 masks soon after. This all happened nearly a week before any masks were made available by the LA government.[6] In the UK, there are also mask blocs in London, Manchester, Sheffield, Edinburgh, Glasgow, and counting, as well as a mask bloc in Ireland.

The AIDS Coalition to Unleash Power (ACT UP) was formed in the 1980s. Alongside their protests and direct actions to spread awareness about HIV and AIDS, they also pressured the government and drug-testing companies for affordable and accessible health care measures, reinforcing people with HIV and AIDs as the experts.[7] Their Treatment and Data Committee also studied and learned from scientific journals, public health clinical trials and experimental treatments,[8] translating the information in more publicly accessible ways so the information no longer remained in the domain of institutions.

Knowledge is power. ACT UP, and the many activists before and after them, knew how often timelines set by governments and other official institutions moved slowly – and all too often at the expense of too many of our lives.

This is something that the Black Panther Party also knew. Within their sixty-five programmes,[9] they set up free health clinics, run by volunteers and doctors, where people could go for regular check-ups, immunisations and high-quality health care, free of charge. They also set up a foundation dedicated to researching a cure for sickle-cell anaemia.[10] In a time when many of us struggle to see a doctor straight away due to long waiting lists, or feel dismissed during appointments, having other means of health care on a community level could be life-changing.

How can we set up other, more transformative means of caring for each other amid a deep absence of societal care? I speak to Hazel Sealeaf, writer and disability lead at MAIA, an artist-led social justice organisation in Birmingham, with a mission to grow and centre Black artistry and imagination for our collective liberation.[11]

Hazel tells me about the organisation's plans for Abuelos, a fully accessible community centre for artists and the local community. Hazel explains that the name Abuelos, meaning grandparents in Spanish, takes inspiration from the radical hospitality you might experience at an elder's house, or a from a person in the community, where the door is always open and you're 'held and fed'.

While MAIA is still in a 'dreaming space' for what this building might look like, in Hazel's words, 'It will be the physical site of the ideas, practices and principles we care about.' Given the importance of disability justice at the centre of MAIA's practice, they're keen to ensure it's a fully accessible, pandemic-resistant site, with natural ventilation systems, to help air flow in spaces, and temperature regulation. MAIA also plans to do this work alongside a paid community steering group of people representing a range of experiences and disabilities to help design the space.

Currently, they're also launching an Objects and Knowledge Library, which will eventually be housed in Abuelos. The library will offer adaptive technologies to loan in northwest Birmingham, where there's a predominately Black community. Available items will range from ear defenders, portable ramps, walking aids and portable wheelchairs to induction loops. Accessible cooking equipment and adapted bikes may also be available, and people will be able to request objects that will then be bought within MAIA's three-month purchase cycle.

Key to the library will be the space to share knowledge, so there'll also be a dictaphone station, where disabled people can share personal stories and experiences of how to protest, navigate the benefits system and stand up for their rights.

In order to prioritise disability justice in our lives, our movements and communities need to better understand that thinking seriously about disability isn't a dragging obligation, but rather an opportunity – to transform how we live in ways that suit all the versions of us – present and future. We shouldn't be taking any cues from the same state powers that are destroying us, but rather cultivating softer, more compassionate ways of living that start with us.

5.

Migration

5.1: *The empire within us*

. . . how will I reconcile myself? . . . the settler and the Indigenous

Jónína Kirton[1]

When my parents first came from Nigeria to the UK, they were only supposed to stay for a year. Before I was born, when my sister was a toddler, my parents used to tell her that by the time she grew tall enough to touch the door handle she was so desperately reaching for, we'd be back home in Nigeria. Fast forward through ongoing political instabilities in Nigeria, more children, career progression and raising a family that would gradually become more integrated into the UK. I live as a product of these choices – writing to you in English, having only visited Nigeria a handful of times.

My family may have stayed in the UK, but the idea of home remained elsewhere, a floating, transitory idea. Nigeria, known growing up as *Home*, was a place of salvation, of return. It was in the boxes of things to be stored and sent back ahead of our eventual homecoming. Over time, as I grew older, *Home* became a disciplinarian – something to be feared when you were in trouble. There was never a low exam grade or illicit eyeliner to be found without the threat of being sent *Home* thrown in the mix. It was supposed to straighten you out, a cure for whatever ills you'd been infected with here. These were just a few of the many instances when I felt the tension of those unexpected plans to stay in the UK as I grew into someone other than Nigerian. Sometimes, despite the choices my parents made, I'd see the disbelief of it written across their faces. *Who is this child that speaks my oppressor's language . . . in a northern accent?*

Clearly, life doesn't always pan out in the ways we think it will.

To be of the Black diaspora living in the imperial core is, in many ways, like trying to love two sides at war with themselves – embodying conflicting signals and inescapable trigger points. Colonial violence and war, historical trauma, migration, disrupted cultural and spiritual belonging – all live within my Nigerian-Britishness. Shaped by both, but never really belonging to either. It's perhaps why I have always loved travel and living in different countries. Because it's where you are called to reintroduce yourself all over again, sometimes in more curious company, without the baggage of either identifier. Here is the chance to look at yourself with some distance: *Who am I now? What new name should I find for myself?*

There are also drawbacks – or rather, more points of confusion, more layers to unveil. When travelling internationally for work, the allegiance to nationality, and the need to pin you to somewhere, can feel even more pronounced. I remember this particularly at a conference in Mexico City, where I was fortunate enough to be performing with a number of poets from around the world. The compere, in trying to raise the energy of the room, introduced everyone and the countries we were representing.

'Give up it up for South Africa!' he yelled, met by whoops and cheers.

'Let's hear it for Uganda!' to which the crowd gave growing applause!

'And ... the UK ...' Cue tumbleweed, followed by stilted clapping.

In that moment I wanted to say: *Excuse me?! You see I'm Black, right?! Consider my struggle! I feel like a clown to be representing a country that shows the countless ways it doesn't want me. Can't I be British-adjacent, without its violence? Without whatever it means to the rest of the world? This room?*

But perhaps this – wanting a piecemeal and denialist version of Britishness – is one of the most British things I could ask for.

Ouch.

The waning popularity of the UK across the world is growingly apparent, and more broadly, we are witnessing the steady decline of Western imperial rule. These are powers that established the global pecking order, giving rise to obnoxious terms like 'third world' countries and the 'developing world', while reigning triumphant over all of it – never once conceding to the fact that they were the cause of said development needed in the first place.

These forces are crumbling, both slowly and quickly. We can see this in the rise of global powers such as BRICS, a group initially comprising Brazil, Russia, India, China and South Africa, which were joined in 2023 by Saudi Arabia, Iran, Ethiopia, the United Arab Emirates, Egypt and Argentina.[2]

West African nations are leading a sea change when it comes to unpicking their colonial ties with France. Mali has dropped French as its official language, following a referendum where 97 per cent of Malians voted in favour.[3] As a nation with one of Africa's biggest gold reserves, Mali is also clamping down on foreign mining companies operating in the country.[4]

Senegal's prime minister, Ousmane Sonko, has called for a reassessment of ties with their former coloniser France, and for 'a change in historical patterns'.[5] The country's president, Bassirou Diomaye Faye, has since called for the French military to close their bases and leave the country, a presence which, he stresses, contradicts their country's independence.[6] Côte D'Ivoire and Chad have also made similar plans,[7] with Chad stating that this is the right time for the country to 'assert its full sovereignty'.[8]

In response to these changes, French president Emmanuel Macron, in a stunning diplomatic blunder, claimed these countries were ungrateful and had forgotten to say thank you.[9] Though Macron conveniently fails to acknowledge the high levels of wealth and resources France has extracted from these

same countries through its colonial exploitation – resources
that enabled France to have a continued military presence
there in the first place.

The backlash against Macron's speech has been pleasing
to watch. Burkina Faso's leader, Captain Ibrahim Traoré,
denounced Macron's comments as 'insulting to all Africans',
adding, 'If anyone is ungrateful, it is him. France exists today
thanks to our ancestors.'[10]

Traoré, a rising figure of the African decolonial movement,
has rejected international financial aid and has prioritised
strengthening his country's infrastructure in his first two years
in power.[11] Traoré has created new roads, as well as upgraded
old ones, has built a new airport and brought MPs salaries
down by 30 per cent.[12] He has also banned colonial style wigs
in his country's courts.[13]

The role that governments such as the UK and US play in wars
and genocides across the world is still happening, but with a
raised public consciousness that is loudly critiquing and protesting these powers. With the horrors happening in Palestine
being the most documented genocide we've seen in modern
history,[14] it's clear that witnessing this has left many of us
deeply changed and ripe for resisting the influence of imperial
powers in the world – a shift that is only set to continue.

The hold the British monarchy once had over much of the world
is also weakening, symbolising more of an outdated regime
than anything of more meaningful value. Barbados removed
the Queen as head of its state in 2021,[15] and a number of Caribbean nations, including Jamaica, Belize and the Bahamas are

planning to follow suit with the King.[16] During a royal visit to Australia, Indigenous senator Lidia Thorpe shouted, 'This is not your land, you are not my King.'[17] A feeling that's also being increasingly shared by the British, particularly among the younger generation, for whom the Royal Family holds little relevance.[18]

Culturally, too, the reverence for Western cultures is wearing thin. The American Dream is being increasingly debunked by the country's volatile political system, extreme levels of poverty and wealth, and its continued oppression of Indigenous communities and instances of racial violence. Not to mention the pompousness it takes to self-promote every US president as the 'leader of the free world', a catchphrase which populated every US and UK political drama series in the 1990s and 2000s.

This is a status only achieved through some of the most vicious, genocidal acts in history across the world. But even in the US, it's hard to see this 'free world' within its unspeakable levels of gun violence and its immoral health care system, where sick people without health insurance can be left to die and abortions, even miscarriages, are being increasingly criminalised.[19] Whose freedom and at what cost unsurprisingly never made the byline. But through the eyes of an evolving, more culturally and politically literate world, it all just feels pretty garish.

Now, as the US is terrifyingly in the throes of a deeply fascist regime, we are witnessing the regressive rolling back of rights, with the pretence of freedom long gone, as the country uses the same heavy arm of oppression it has cast to others on its own people. The US is imploding, with the Republican administration, and Trump's tech-bro friends, gutting its public infrastructure, dismantling the department of education[20] and cutting social welfare programs, including social security, disaster relief and affordable housing projects.[21]

Far from being the self-appointed world mediator it used to be – making itself both judge and juror of global discussion – the US is now a temperamental toddler on the world stage, with overt hostility towards its former allies and seemingly no attempt at political diplomacy. How strangely apt that the downfall of this global superpower should come from none other than the country itself.

Of course, over the years I've had conversations with others about what kind of future there can be for Black people in the imperial core. Where do we fall in with all the global changes happening around us? Certainly, in the UK, the steady rise of fascism and right-wing extremism has intensified our own struggles and more brazen levels of aggression and racism – if not in action, then in the very tangible fear of it. The veneer of racial progress is wearing thin. Given many of our family histories, I think conversations around leaving the UK honour a legacy many of us were born into. Our parents moved here for a better life, at great cost. If we too find ourselves asking ourselves similar questions about the future, isn't it our birth right to do the same thing?

If one of those questions may be whether to go back 'home', then let's be honest about what we risk taking back with us. For many of the Black diaspora, we've been raised and educated in institutions of oppressive thinking masquerading as enlightenment.

To survive here, we've had to learn colonial languages, knowledge and its ways of thinking by heart. The thinkers, scientists and intellectuals we've had to learn from have been steeped in Eurocentric ideology. And we've internalised racism just as keenly in the absence of what hasn't been said – in truths rarely acknowledged, and histories and movements never studied. The assimilation – changing our names, having our cultures and family members' accents mocked – is scarring. (I once kissed my teeth at school, and the girl sitting next to me thought I was trying to kiss her – I was six.)

We've been taught to forget ourselves. Some of us have, completely. More than we can know, on a conscious and subconscious level, we've been fundamentally shaped by living within an imperialist mindset. And as problematic as we know this to be, to live here and even attempt to thrive, we've had to buy into its logic and play by its rules.

I know life is rarely ever this straightforward – not every family who came here settled. Those that did may well have managed to combine and balance both cultures, still maintaining *Home* in their countries of origin. For many African Americans, transatlantic slavery has snatched the idea of 'home' beyond the US completely. Others among us will have fled from persecution and perhaps don't have those same communities or even land to return to safely. Decolonial movements have also offered us ways to deconstruct the colonial programming we've been raised in. But still, I can't help but wonder how living in the motherland of empire has fundamentally changed us – in ways beyond the obvious, quietly and without our knowing.

I'm often heartbroken by the hardness I experience when returning to Nigeria. It took me a while to realise that my parents' home wasn't my home – at least, not in the same way. There's often a specific and unspoken feeling in being there that can be hard to put into words – but it's a sense of heavy alienation. I remember, especially when I was young, trying to at least blend into the background – but my clothes, demeanour, or – the real clincher – my inability to speak Igbo, always made me stand out from a mile off.

In getting older, I can see the other side of this. What it must be like for people who stayed to watch their neighbours, friends or family members make decisions you didn't – to go and live in the backyard of an imperialist country that has done so much harm. To watch those same people come back, more changed with every visit, and wealthier. While people of the diaspora are much more likely to bring wealth back into their countries of origin, through investment and sending money, it could also be a strained relationship to be on the receiving end of it.

Those who stayed perhaps have a sharper view of colonialism – in living within both the environmental and political instabilities inflicted on their countries as a result, but also in what our absence says about the country they've stayed in. I see that judgement in their eyes, even if I'm more of a consequence of the decision to leave. The suspicion that I, too, must see them as lacking, as if through the eyes of the coloniser. That I could also commit the same levels of harm, as history has shown the *oyinbo** is prone to do.

* A word used for white people or westerners in general.

What harm do we have the potential to replicate in our return? And through whose eyes do you see your country(/ies) of origin? Do you believe you can 'fix' or 'improve' things? If life became uncomfortable – dangerous even, particularly with the sharp end of the climate crisis happening across Africa and the Caribbean – how soon would you pull out your other passport and leave? As the global power of the imperial core shifts and dwindles, will you, in our own small way, feel the loss of the privileges living amongst those same powers once afforded you, given all that was lost to get there?

This is not to undermine those who find home in their countries of origin, or even any other Black-majority nation, but rather to highlight some of the contractions we risk arriving with. The Ghanaian government's 'Year of Return' initiative was launched in 2019 to attract more people of the African diaspora to relocate there.[22] But even that offer hasn't been without its complications. Five thousand acres of land were put aside as plots for people to build on, an offer that has largely been taken up by people from the US, Canada and the Caribbean.[23]

But this has come at great expense to the 150 farmers who actually own this land, none of whom have consented to these changes, or have been compensated, particularly given how this displacement affects their capacity to grow and sell produce from the land for their livelihoods.[24] The idea that this space was 'freely available', despite the actual reality, is a colonial attitude made much more uncomfortable by the fact that it's now happening because of the Black diaspora. This is a violence that has been learned, passed on.

The nostalgia of return is a romanticised myth. A new start, yes, perhaps even meaningful. But it should come with none of the assumptions our indoctrination of Western ideologies might lead us to believe about what we are owed. In her book *Lose Your Mother*, the US academic and writer Saidiya Hartman reflects on her time in Ghana, where she explores her imagined ancestry and personal histories in the aftermath of slavery. She describes trying to find belonging but is disillusioned by the reality and the way she's mocked and viewed by her African colleagues. She writes:

> The country in which you disembark is never the country in which you have dreamed. The disappointment was inevitable . . . was it foolish to long for a territory for which you could risk imagining a future that didn't replicate the defeats of the present?[25]

Wherever we go, we take ourselves – our mindset, our assumptions and our expectations. Much of these have been informed, at least in some ways, by living under empire. Global unrest is also catching; protests and political frictions are happening all over the world and across the African continent, with Nigeria, Kenya, Zimbabwe all respectively fighting against the violent suppression of democracy. Layered in with the climate crisis showing up in varying ways across the world, there is no utopia and no return that solves the issues that are worrying us in the here and now.

As Hartman adds:

> The hope is return could resolve old dilemmas, make a victory out of defeat, and engender a new world order. And the

disappointment is that there is no going back to a former condition. Loss remakes you. Return is as much about the world
to which you no longer belong, as it is about the one you are
yet to make a home.[26]

Loss remakes you, and with this comes both the responsibility
and freedom to reimagine where and how we seek home.

5.2: A question of belonging

The West has imposed two fears on itself: terrorism and migration

Aminata Dramane Traoré[1]

CW: racial violence, xenophobia, anti-immigration rhetoric

Migration is complicated and is tied up in a racist discourse around who gets to belong where. As I write this, white terrorism and race riots are unfurling across the UK, sparked by the tragedy of the murder of three girls in Southport by a Welsh teenager of Rwandan heritage. *#EnoughisEnough* was the hashtag many far-right protesters took to using on X (formerly Twitter), declaring that this was the last straw. Despite the fact that the young man in question wasn't an immigrant, his ethnicity was fuel enough.

This is one of the most visceral waves of racial violence and hatred the country has witnessed in decades: acid bombs, stabbings, the chasing down of Black and Brown people, strangers pulling Muslim women's hijabs off and targeting mosques, starting a fire in a hotel where asylum seekers are housed and threatening to target community support centres for refugee and asylum seekers.

The logic behind this has been patchy, to say the least: immigrants are supposedly stealing all the jobs but are also lazy and unemployed. Immigrants are dangerous and a threat to the community, yet it's rioters smashing shop windows, attempting to set buildings on fire and devastating local communities.

There have been the predictably classist takes as a response – the mocking insinuation that this is solely a white working-class problem. While these are some of the most shocking outbursts of violence I have witnessed in the UK, they're also entirely familiar. Black people have also experienced this violence in our places of work and in mainstream media, from white friends, partners and health care professionals.

Every white person has the potential for this level of racial violence within them. It's an inheritance. Unless they have spoken back to its inherent entitlement and endless belief in its goodness and supremacy, there will always be a breaking point. It might not look like burning hotels or looting a Greggs, but too many of us have been there when the mask of pretence eventually slips and we experience an anger just as vicious.

The riots of 2024 are a prime example of whiteness in tension with class struggle. As life gets harder, and the UK loses its power on the global main stage, those among the most disenfranchised are realising the carrot on the stick leads nowhere. Their whiteness has sold them a lie and has given them an entitlement that has produced very little. This has been underpinned by political elites who have told them the reason for this is the migrant, standing in the way of the wealth and power they're owed. The good life, untouched by difficulty or crisis.

For the rioters, and more broadly those who embody this kind of anger, it's not that the order of things is necessarily wrong, but rather their disbelief that they're stuck in the struggle with the rest of us – something that their whiteness has told them should never be possible. It's this pride that stops us from having a much larger level of class consciousness in the UK where we, as a collective, can come to understand more of the similarities in our oppression than our differences.

The looting is only a mirror image of what whiteness does – take without asking and destroy without consequence. In this case, it's small-scale compared to what has been taken from the world. But really, it's a rejection of the idea that they have more in common with the rest of us, and a distraction from a far bigger truth.

The riots were an eruption of a manufactured anger, built on years of sensationalist mistruths about immigrants, which have been widely platformed by our politicians, extremist public figures and the mainstream media. Immigrants have long been a handy scapegoat for government failures. The lack of jobs, the cost-of-living and housing crises, and the lack of sense of safety and community – despite any evidence, rhyme or reason – are all blamed on immigration.

For decades we've been inundated with anti-immigrant hate speech. Former Tory Prime Minster Theresa May famously boasted about creating a 'hostile environment' – a series of policies and measures so difficult, it would encourage 'illegals' to leave.

It's this sentiment that would have the government justify the subpar standards many asylum seekers face when they arrive here. With many likely to be placed in temporary accommodation or a hotel anywhere in the country – which they have next to no choice over – until their asylum case is determined, which could take years. The hotels can be cramped and dirty, which is especially hard for disabled people and families to navigate.[2] Asylum seekers are given what's called an ASPEN card, a prepaid debit card with an allowance of £8.86 a week per person.[3] While they're given food, it's often of poor quality, sometimes even mouldy.[4]

Even in cases where migrants have all the necessary documentation to live and work in the UK, many are refused housing, employment and health care.[5] These realities are a far cry from the fairytales of foreigners coming into the country for a 'free meal ticket'. And yet the mistruths around this continue.

The disaster of Brexit was pitched as a way for us to rid the country of immigration, but that still didn't stop us from being told, some years later, that a 'hurricane of mass immigration is upon us'.[6] Raids, deportation centres and the deplorable plans to deport people to Rwanda have all been implemented and proposed under Tory rule. Rishi Sunak even had 'Stop the Boats' plastered on his podium during an immigration speech,[7] which many of the rioters subsequently took to using as a chant. The connections are crystal clear.

While some may feel relieved that Labour Prime Minster Keir Starmer has scrapped the Rwanda plan, the Labour Party has also contributed to anti-immigration sentiment, never truly offering an alternative line or approach to the situation, other than framing immigration as a problem in need of solution. In fact, Starmer's shocking speech in May 2025 in which he argued that we risk becoming 'an island of strangers' without more hard line policies around immigration,[8] was a dangerous echo of the racist anti-immigration rhetoric in the country- of the past and present.

Anti-immigration rhetoric is so effective because it builds on long-standing racism, mixed with policing, control via borders and nationalism. It raises questions of who gets to live here – who is deserving and who is not? Whose movements should be monitored and policed?

Race and desirability are central to this, which is why white Irish, Canadian, Australian, Ukrainian and other white European immigrants are rarely included in these debates. Nor are the Brits who proudly call themselves 'expats' and who rarely make efforts to assimilate when they emigrate to other countries. It's also why when extremist Tommy Robinson (real name Stephen Christopher Yaxley-Lennon) was arrested – ironically for being in Canada illegally – his response in the back of the police car was, 'Innit mad* how tough you get with immigration on the wrong people.'[9] Whiteness always expects to get a pass.

Contrary to what we're told to believe, migration is a good thing. According to the Office for National Statistics, it brings £83 billion to the economy a year,[10] which is far more revenue to the country than it takes out.[11] For all the racist scaremongering about a 'hurricane of mass migration', in reality immigrants make up roughly 14 per cent of the population,[12] which is predicted to drop given UK policy changes to visas and an increase in the minimum salary people have to earn

---

* An ableist term, but taken from a direct quote.

to enter the country.[13] The UK also receives far fewer asylum applications in comparison to its European counterparts[14] and general attitudes towards migration are more positive than they've been in recent decades.[15]

While there's a large body of evidence that shows how migrants are good for the economy, migration is a fact of life with inherent value in its own right, not just in ways migrants are in service to the country. I think we're often pushed to think about migration in terms of figures and cost benefits because of how twisted narratives about the threats of migration have become. But to think about migration in this way alone is dehumanising. Morally, why shouldn't we all be able to travel anywhere?

The term 'strong passport' describes those who can travel to most countries without a visa (Singapore, Japan, several countries in the EU and the UK are on this list) and yet there is far less of a moral outcry against this. I get this is because of the perceived economic benefits people from those countries might bring. But given the way imperial countries have destabilised countries across the globe, isn't freedom of movement the least people from those countries are owed? Not to mention the deep cultural benefits migration offers to the UK alone. This country's food, music, literature, fashion and cultural landscape would be significantly poorer without it.

It's important that we're able to shift though the noise of a small but loud minority when we come to understand migration. We must be on the better, more flexible side of change, particularly as the realities of the climate crisis may intensify migration, reshaping our relationships with movement worldwide.

5.3: The other side of the border

CW: climate disaster

Noora Firaq, former Deputy CEO of the charity Climate Outreach, was born and raised in the Maldives, a small island nation southwest of India and Sri Lanka.[1] Growing up, Noora remembers it as a peaceful country that changed from a fishing economy to one largely reliant on tourism. It's also a nation that's extremely vulnerable to the climate crisis, and in Noora's words is 'without a single hill, mountain, river or nature reserve'. As the world's lowest-lying country, 80 per cent of it is less than one metre above sea level.[2] While flooding and rising sea levels should be a concern for all of us, for a country like the Maldives, it's nothing short of devastating.

You'd think that this might be the reason Noora moved. But she originally came to the UK for university, and for a long time, that was her primary reason for leaving her country. It took her fifteen years to realise there was a climate dimension to her migration story, which begins to paint a far more complicated picture about the climate crisis and why people move. In her words:

> We often frame the climate crisis by rising sea levels, a loss of ecology, or not enough fish in the sea, but thinking of my country's trajectory made me realise when you spend most of your budget on climate adaptation – building sea walls, fighting against sea erosion – you end up not investing enough in your future; in educational systems and health care.

These are the secondary effects of climate change, where countries that are particularly vulnerable have to invest much more in climate adaptation at the expense of other public investments that might be needed.

So, for Noora, moving was less because the Maldives were sinking, and more because the education sector was under-resourced, making it hard to access a good level of education, especially in rural areas. Reliable health care is also patchy, with many having to go to nearby countries, like Sri Lanka, for more complex surgeries. This has increasingly turned reliable health care into a privilege that not everyone can afford. These are the ripple effects of the climate crisis, felt as keenly in what doesn't get funded as in what does.

The climate crisis stands to change many of our environments, and with it, worsen existing climate vulnerabilities experienced across the world. As a result, we're likely to experience much higher levels of global migration in our lifetime. While narratives around any kind of migration may well be largely dominated by 'foreigners' coming to the UK, the reality is that the climate crisis is likely going to change all of our relationships to place and migration. According to The Migration Policy Institute, 3.2 million people were displaced or evacuated in the US due to natural disasters in 2022 alone, and the government has already started to relocate communities vulnerable to rising sea levels.[3]

The LA wildfires at the start of 2025 are also a shocking example of how environmental disasters can displace people, either temporarily or, for those whose homes were destroyed, for a more long term. While there may be a sense of relative safety for those of us living on the cooler side of the hemisphere, it can't be separated from the arrogance and untouchability often baked into imperial cultures. Because the reality is we will all be affected – albeit in different ways.

The entire community of Fairbourne, a seaside village in Wales, is due to be relocated and the village 'returned to the sea', given its proximity to the rising levels of water year after year.[4] Similar questions and strategies on how to protect communities are likely to be developed in many coastal towns across the country, such as Norfolk, Devon and Cornwall.[5]

Lancaster is likely to suffer from significant water shortages by 2040, as will Birmingham, Leicester and Northampton.[6] London will also be vulnerable to more flooding as the Thames rises within the next ten years, affecting much of west and east London.[7]

The increased frequency of flash flooding and heatwaves may be unbearable for many of us, and will be an added strain on our public services. The heatwaves of 2022 were the fire brigade's busiest time since the Second World War, when temperatures of up to 40°C saw over a thousand fire-related events in London.[8] The flash floods in London just a year before in 2021 left cars and buses stranded, Tube stations closed and over 300 calls were made to the fire brigade in a single afternoon as homes were flooded across several parts of London.[9]

When we hear these stories from a safe distance, it's easy to think of the event alone, but days, weeks, months – even years – afterwards, people continue to live with the traumatic aftermath, particularly where there is injury and damage to people's bodies and homes.

As these incidences increase in intensity, it may well be necessary and lifesaving for some of us to move to different parts of the country that are less affected. This would have to be done with support and intention, particularly for vulnerable and disabled communities, who need accessible, often specialised, medical care.

That said, the largest patterns of climate migration are more likely to happen in other parts of the world, in Africa and the Asia-Pacific region,[10] where many cultures rely on agriculture, crops and animals, all of which will suffer directly under increasingly extreme weather conditions.

Movement in response to a climate crisis brings its own set of complications; whether as a rapid response to a disaster that happens quickly, such as a flash flood or a typhoon, or slowly via drought or rising sea levels.[11]

While the rapid-onset disaster may be over in a matter of hours, people are usually less likely to be prepared to move. The aftermath will also be felt for much longer, whether though damaged infrastructure, the possible death or injury of loved ones, changed ecosystems and the increased risk of disease – not to mention shock. Patterns of migration around this may be more temporary, in a safe place nearby, while communities are able to rebuild or return to their homes.[12]

Slower-onset disasters raise more existential questions and are more likely to lead to permanent moves. What choice do people have living in, for example, Tuvalu – an island in the Pacific Ocean projected to be underwater within fifty to a hundred years?[13] Or the 30 million people in Nigeria, Niger, Chad and Cameroon, relying on Lake Chad, which has shrunk significantly over the past sixty years?[14] These are people's livelihoods, basic needs and communities all being jeopardised by the changing of their worlds as they know them.

Movement in these instances is likely to be due to several connected factors – not just environmental. Other pressures, such as a government's inability to respond to climate threats, or as in Noora's case, a lack of public infrastructure, also compound the issue. Displacement might also occur because of the rising tensions that come with scarce resources, such as conflict and war.

Climate catastrophes across the world show how the cruel legacies of colonialism still reinforce themselves today, and maintain the global inequalities they created, centuries later. This is because the culture of capitalism – violence, extraction and hoarding extreme wealth – took root under colonialism. Land and ecosystems that had long been sustained by Indigenous communities before were destroyed by imperial powers in the pursuit of profits that those exploited communities never benefitted from.

It also led to the normalisation of unsustainable ways of operating, of taking without any view of the damage being left behind, or any thought for future generations. This, in the process, eradicated Indigenous practices that had been developed in connection with the seasons and the land, instead paving the way for mass production, pollution and overconsumption.[15]

In the words of author and campaigner Daniel Macmillen Voskoboynik:

> Nature narrates the colonial story, through its vast mines, its desecrated rivers, and emaciated territories. Across continents, mangroves, grasslands, rainforests, and wetlands were cleared to make way for quarries, plantations, ranches, roads and railways.[16]

It makes me think of Sierra Madre in the Philippines, one of the largest and most expansive mountain ranges in the country. It's a natural line of defence for the Cagayan Valley region,[17] known for its capacity to shoulder heavy winds, storms and tropical cyclones.[18] It's an ecosystem that also regulates extreme weather conditions due to its capacity to absorb the carbon dioxide from the atmosphere.[19] And yet, 90 per cent of its rainforests are gone due to deforestation, illegal logging and mining – a consequence of the area's rich reserves of copper, gold and limestone.[20]

Our natural world has its own intelligence, not to mention in many cultures spiritual and cultural significance. The consistent pillaging of these natural resources is quite literately leaving many countries defenceless. The deforestation of Sierra Madre has led to more flash floods and landslides due to fewer trees being able to absorb the water,[21] not to mention the loss of biodiversity. Tellingly, illegal deforestation was set in motion during the Philippine's 333-year colonial history with Spain, in which period, 25 per cent of its forests were cleared for economic gain.[22]

This culture of resource extraction under colonialism has taught the world which ecosystems are ripe for exploitation, and has left those same parts of the world with lasting difficulties in trying to navigate the climate crisis. We can only imagine what other natural worlds have been destroyed by colonial exploitation – landscapes that, by their design, once offered sustenance, shelter and natural protection against more extreme weather conditions. Now gone, it leaves much of the world vulnerable and out of sync with the design and structures of their natural habitats. As the organiser and campaigner Dorothy Guerrero writes:

> Most countries with the highest climate vulnerability and those most vulnerable to climatic destabilisation were formerly colonised countries.[23]

Migration is just one way in which we can see the aftershocks of such violence. If not because of increasingly unliveable environments, set in motion by imperial exploitation, then because of the ongoing political and economic instabilities many countries still struggle with in its aftermath.

Noora reflects on changes across the years of growing up in the Maldives – a former British colony – and now on her yearly visits. Diets are different now. There are more land disputes, as less is available and often divided up between family members, if at all. She also worries about the growing rates of murder, violence and radicalisation, something she remembers as rare in her youth.

Resistance to migration and asylum, particularly within the imperial core, is a resistance to the consequences of the violence those same imperial players perpetrated. Perhaps fewer people would want to come to countries within the imperial core if their countries hadn't been so irreparably changed and exploited. With this understanding, the UK's 'hostile environment' policy is a particularly biting idea, given the hostile environments the UK has also played a huge role in creating within the many of the countries people are fleeing from.

To be of the Black diaspora in the imperial core is to live between these realities, living within our own hostile environments, and the violence of racism. This happens simultaneously while being expected to live within a sanitised version of events, which would have us believe none of this is happening. There's a reason borders matter in the imaginary of the imperial core. In enacting them, the atrocities that have been committed can be kept contained, relegated to a place out of sight and out of mind.

Communities, movements and activists across the world have been campaigning for decades to reduce emissions, calling on global powers to act in ways that heal the planet rather than harm it. Refusal to do this in any meaningful way is largely because for so long the climate crisis has been understood as something that happens on the other side of the border, which imperial powers are committed to enforcing at all costs. This continues a legacy many are very comfortable with – the ongoing suffering and pain of non-white bodies. The idea of open borders, or at least freer movement, blurs this distinction, as does the very nature of the climate crisis, as islands, ecosystems and biodiverse plant and wildlife worldwide disappear because of it.

If the climate crisis can teach us anything, it's that borders are an illusion. While we can't deny the disproportionate harm happening in specific parts of the world, we'll all be affected in some shape or form.

Climate change calls on us to look at the world, beyond its borders, and offer solutions that are much more centred around tackling the problems we face at hand. This has to be beyond old, entrenched patterns of division and resource exploitation, which make no sense in the face of a crisis that needs our solidarity, collaboration and collective thinking. This is the only planet we live on, and our solutions need to be planet-centred, not the same old dusty attempts to hoard control and power.

We need solutions beyond borders, but for everyone, everywhere.

5.4: Climate migration – A new reality?

CW: carceral violence, gross human rights violations

As someone who's second generation, born in the UK to immigrant parents, I've always been aware of the fickle nature of migration and borders – a trick of time, place and political will. Had my parents decided to move to the UK any time before and after their actual arrival, it could have just as easily led to a completely different set of experiences for my family, purely because of a set of policies that happened to be in or out of place during that time.

This is why I come back to migration a lot, especially in my poetry. I'm so aware of the illusion of it – that there's no one more inherently deserving than anyone else trying to move here, and that we're all subject to the fluke of our circumstances. The line of acceptability when it comes to migration is moveable and is constantly showing its inconsistent and changeable nature.

Seeing the 2024 race riots in Rotherham, where I grew up, and the hotel where some asylum seekers were staying be set on fire, I'm especially aware that the only difference between me and those asylum seekers is luck. This is also why conversations around immigration that only focus on 'illegal' migration are transparently racist. It's a thinly veiled suspicion of all of us, and that suspicion doesn't go away with a visa or a UK passport.

We're seeing the sharp end of this in the US where hundreds of Venezuelan migrants are being sent to CECOT, a mega-prison in El Salvador, without trial or any legal procedure,[1] which in itself is a serious human rights violation. Happening under the guise of tackling 'illegal immigration' and gang violence, emerging reports reveal many of the people detained were in the US legally and have no criminal record. ICE agents are even questioning, and at times imprisoning, people on the basis of their tattoos.[2]

The prison, well known for its torture and food deprivation,[3] runs under strict conditions. People imprisoned there are never allowed outside, nor any visitors.[4] They've also been stripped down and put into uniforms, cuffed both at their arms and legs, and their heads have been shaved. These dehumanising images are terrifyingly reminiscent of concentration camps, something that the Republican administration seem to be proud of. US Secretary of Homeland Security, Kristi Noem, is even seen in a video standing outside one of the prison cells near boasting that this as one of the consequences of illegal immigration to the US.[5]

We are living through a harrowing time of anti-migration sentiment, and it exists as a continued form of racial violence emboldened by the power of national security. The way the parameters around who's considered undesirable continue to widen should be a grave concern to all of us.

Such tensions and, in many cases, violence around migration have created an inwardness – so dense that we are missing a much larger conversation around the mass migration the climate crisis is likely to cause. While countries within the imperial core are scrambling around and avoiding the truth of their national identity, a much bigger issue awaits – that of migration on a global scale, which will require a global adaptability and pragmatism that is currently sorely lacking.

Displacement and migration from the climate crisis are likely to escalate, particularly in parts of the world that become uninhabitable due to sea level rise, desertification and extreme weather conditions. Opinion on the scale of migration is varied and complicated because, essentially, we're trying to predict human behaviour. While some have predicted that we could be seeing over a billion people move due to climate migration,[6] it's a figure that has since been debunked and largely problematised.[7] Other predictions are more conservative,[8] given that people are much more likely to want to stay in their homes, or at least as close to them as possible.

This mirrors current patterns of migration, where a large proportion of people choose to migrate internally or to neighbouring countries, rather than across continents. The future of climate migration will also depend on how countries invest in and adapt to their changing climates and the climate mitigation measures put in place.[9]

Sensationalist claims that don't consider this and get swept into the hysteria of 'mass migration' narratives not only fuel much of the far right's paranoia around migration but also reveal a lack of humanity and understanding towards the people they are talking over.

If you had to move at short notice, due to a climate disaster, where would you go first? To friends or family? A designated safe zone in your community? Or halfway across the world? The reluctance and fears *you* would have in leaving – when you otherwise would have wanted to stay home – are the exact same feelings and responses experienced by people who are actually forced to move. Not to mention how expensive a move like that would be for many of us.

We have to give room for this and understand some of the real human complexities of migration rather than fall back on these one-dimensional narratives. Noora also confirms this, stressing the real need for us to develop an alternative language around climate migration. In her words:

> When there's a climate disaster, people don't have time to look at a map and go to the UK, they go to the nearest place where there's safe water.

What remains true is that the climate crisis will reconfigure which parts of the planet are habitable, with some tropical regions likely to become too dangerous for human habitation. Colder climates are also warming. Siberia, known for its grizzly winters and below-freezing temperatures, has been experiencing a number of heat-waves[10] and is, according to the World Meteorological Organization, 'one of the fastest warming regions on the planet'.[11] Canada's Artic region has also been experiencing heatwaves of up to 33°C.[12]

With such impending change on the planet, growing trends and threads of thought are emerging. Gaia Vince, author of *Nomad Century: How Climate Migration will Reshape Our World,* believes that climate migration will be one of the defining issues of the century. She argues that those of us living in cooler parts of the world will have to accommodate those who will be displaced, which could see a reconfiguration of the make-up of our cities across the world. She also highlights how countries in North America, parts of Europe, Japan and the UK have ageing populations, and so will need the benefits of migration. Meanwhile, many countries, such as those in South Asia or parts of Africa, where mass emigration might be likely due to the climate crisis, have a majority younger population who will be of working age.[13]

There's also a shortage of people skilled in green jobs,[14] which includes developing and implementing green technologies such as solar panels, wind power and improving our existing buildings to make them more energy efficient. The Boston Consulting Group predicts that we will have a global green skills gap of 7 million by 2030.[15] Meanwhile, all electricity in Nepal comes from renewable sources,[16] and Kenya also generates around 80 per cent of its electricity from wind, solar, geothermal and hydropower, with a target to fully commit to green energy sources by 2030.[17] Not only will the upskilling of people in the UK be necessary, but the skills and talents of people from countries across the globe will also be very much needed in the future.

Additionally, the UK continues to have a significant need for workers in the health and care sector. Despite Brexit, migration has largely held this system together for years. In 2022–2023 alone the UK saw large numbers of international health and care workers coming to the country.[18] This dependence on migrant workers is likely to continue, especially without significant reinvestment in the NHS and a commitment to address staff shortages.

These shifting labour and migration patterns point to more flexible movement as the climate crisis, among others, calls for us to evolve. However, I'm less drawn to the economic opportunities of these shifts and far more concerned with the potential for the same exploitative patterns to emerge from these new realities. If mass migration is to happen on any large scale within the current system, it risks propelling younger Black and Brown people into the heavy labour that props up the imperialist project – again.

Colonialism across the world funnelled some of the very best minds from colonised countries into servitude, whether working as cheap labour in mines, plantations, factories or domestic labour, set up for colonial gain.[20] People working within bureaucratic positions – administrators, including leaders – were also often in the pockets of colonialist powers, executing their demands either by bribe or threat of death.

Of course this can't be said for everyone – I would never want to be that simplistic, because there have always been people on the side of anti-imperial resistance. Emperor Menelik II and Empress Taytu Betul of Ethiopia, along with an army of 100,000 people of different ethnic and religious groups, resisted Italian colonial forces outright at the end of the nineteenth century.[21] Négritude, a cultural and political movement, started in the 1930s and was led by poets and eventual political leaders Aimé Césaire, Léon Damas and Léopold Sédar Senghor. The movement was a reclamation of Blackness and African culture, and staunchly opposed to colonialism. And the Malian midwife,

politician and independence activist Aoua Kéita was an instru-
mental member of the anti-colonial political party US-RDA.[22]

I also know there are so many more examples that have never
made it to a history book, given just how many of our experiences have been hidden and erased in the pursuit of simplistic
imperial narratives of conquest. I leave space for them here.

But there is no getting around the heavy realisation that many
Black and Brown bodies have been the wheels that have turned
the colonialist machine. In the process of us operating under
their agenda for financial gain, we've lost collective memories
of our histories – perhaps even our purpose.

I think anyone who wants to move should, but there's also
something uneasy about the idea of more of us becoming
further disconnected from our ancestral lands and histories –
and for what? Coming to the UK via unsafe routes of passage,
working jobs that are crucial and yet devalued, many on precarious wages, and having our identities questioned in the
process?

Even for those of us who have made the move, either as a
migrant, or now integrated as second-generation immigrants,
we've largely had our intelligence and creativity frittered away
by a system that takes the best of us yet values us very little.

We've spent decades fighting to be recognised, for equal wages and deserved job promotions, often at the expense of being the leaders we're truly capable of becoming.

The impact we've still had on our cultural landscapes – through art, literature, music and fashion, for example – is testament to our resilience and talents, but these contributions are so often downplayed, or co-opted and attributed to others. Those of us who have made it to positions of leadership have done so under deep scrutiny in circumstances far more difficult than ever necessary.

I've noticed that whenever I've had certain opportunities they have come without a safety net. And any failure, real or perceived, has often been met with the most abrupt backlash, a surge of unspoken anti-Black feeling. It has taken me a long time to move past internalising this (*they were right, and I was so very wrong, just terrible!*) to understand how it really worked. That so many people and organisations, whether they knew it or not, equated my work, even my presence, to inherent risk. So often the opportunities I thought I'd earned were borrowed, given out on a limb, *graciously*, and anything short of perfection rescinded the offer. There is a reason why 'Black excellence' is such a common saying – because there's often so little acceptance of anything else from us. Compare this with the failing upwards we see of many white men in leadership positions, where no mistake sparks any such accountability or consequence.

Do you, we, truly believe in this reality, borne out of whiteness and colonial thinking, enough to advocate we keep adding to it via mass climate migration, not least without deep structural change? While coming to the UK may well offer some people better opportunities, how much does this experience truly cover the compromises and discomforts that sit underneath it?

Without a reckoning with the culture that surrounds migration and the racialised climates we live in, more of us coming to the country are doomed to be stuck in the same fate, and many of us know just how complicated living in the wreckage of that can be.

The system that underpins migration would have to change, and the UK will have to atone for its hostility to The Foreigner, an attitude that feels almost karmic given its own unwanted colonial presence in many countries across history.

When migration becomes an even bigger need and a matter of survival, the state will have to concede to the way it has used immigrants as political scapegoats, the lying and arrogance, the changing goalposts of who can come and stay, the dependence on migration while publicly denouncing it.

Without this, and any deeper consideration, the UK has no business bringing more people into a state where migrants are more likely to encounter carceral cultures, deportation and fewer citizenship rights, that can flex and change government to government.

The migration of the future must be much more human-centred and mutually reciprocal. Nor can it continue to rely on 'good immigrant' narratives, where acceptance is dependent on impossibly high standards of excellence and goodness, expected of immigrants by people rarely achieving this standard themselves. This will be especially important given the slow, and often secondary, reasons for leaving a country – like in Noora's case – which might not be so obviously linked to the climate crisis but should be accepted all the same.

There's also a question of reparations and climate debt. Imperial powers have a responsibility to account for the reasons why climate migration is happening in the first place. When we talk about climate debt, it speaks to the amount that richer countries and their oil, gas and coal companies have gained through resource exploitation, which is said to be a staggering £1.7 trillion a year.[23]

Climate reparations call for these same countries and companies to stop causing harm, account for the damage they've left behind and repay the countries they've exploited. It could also mean supporting these countries to, for example, develop technologies and means to adapt to the climate crisis they are now disproportionately affected by.[24] But it can't come with the same colonial trappings of the present day – in the form of debt, political meddling or any continued 'special relationships' where imperial countries continue to take advantage. I think it also reinforces the need for opening borders if part of the repair means that communities most affected by the crisis seek other parts of the world to live in, should their countries become uninhabitable.

Writing much of this in the wake of the race riots across the UK is jarring, to say the least. Of course I've asked myself: how can I write about open borders and reimagining migration at a time when I'm questioning how safe I am to leave the house? But this is what it means to be living in this time – between ideologies, and worlds dying, with newer worlds to come. It's painful and disorientating. If things weren't changing, there wouldn't be such violent protest and resistance from those hoping to benefit from these systems staying exactly as they are.

If the actions we witnessed from the rioters, but more importantly, the racist behaviours of the political elites, are to be believed, then we have a long way to go before the migration system could be any different. But it's worth remembering that many more of us also protested against the riots and advocated for the value of migration in our communities.

A heartening image that stays in my mind is of the thousands of people who attended the counterprotests to the riots across England – in Brighton, London, Liverpool, Bristol and Birmingham[25]– against a handful of people who showed up to continue the race riots.[26] Migration, for the majority of us, is a good thing. Racial violence and xenophobia can often be overhyped, facilitated online via bots and paid followers who create a reality that – while by no means less terrifying – can barely stand up to the majority of us in the cold light of day.

Freedom of movement may seem like an unrealistic idea, given how our borders are so heavily policed and militarised, but you'd be surprised how quickly things can change, especially when understood to be for the good of the economy. We also have examples of this in our modern history, given that there was freedom of movement in and out of the UK at the beginning of the twentieth century.[27] In our current day, this is precisely what happens in the EU, brought into law in 1992.

The climate crisis forces us to reconceptualise our communities and borders, along with the very real changes to the planet the crisis will bring. But another sticking point I have when it comes to the idea of mass climate migration on the scale some are predicting is that it places the burden on countries and communities across the world that have contributed the least to the crisis, putting them at the mercy of the countries and people they may migrate to.

Also, without the expertise of communities and leaders from countries most vulnerable to the crisis, we're not having the full conversation, nor are we understanding the full extent of what climate resilience can look like. Noora remembers the prediction that the Maldives would be underwater by 2020 and is proud of the ways her country has resisted this. 'We're climate leaders,' she says, adding:

> We are showing there are ways forward, but sometimes it feels like our efforts aren't recognised . . . there will be many stories and communities like this.

She also reflects that this is the kind of pragmatic leadership that is rarely seen on the world stage, or even within the climate movement in the UK. I think this is also part of the language and mindset that need to evolve around climate migration. Who are the real leaders, if not people currently navigating and surviving the worst of the climate crisis right now? What can we learn from their resilience and adaptation, and how can they guide us? Crucially, what do they predict for their own futures?

Moreover, in so many cultures, connection to place and land is sacred, and has been hard fought for against violent colonial forces. This explains why, when movement is necessary, many seek to move as close to their homes as possible, with hopes of one day returning. Many Indigenous cultures know how to tend to the land and its biodiversity and have deep ancestral ties. Historically, many cultures have rituals, ways of mourning and burying their dead that are intricately connected to place and land.

There will be layers of grief embedded in decisions about what to do if it no longer becomes feasible to live in these parts of the world – grief that can't be easily alleviated just by moving. I think perhaps there are lessons the Black and broader diaspora can offer here about what it means to carve a sense of belonging beyond a passport or country to comfortably align with. We've had to build homes and ways of being in the spaces between countries and cultures, to build community and refuge – despite being told, one way or another, that we have no place here. With the climate crisis come new opportunities to imagine our communities in other ways.

Whatever side of the border we find ourselves on, let's understand it's ultimately a fiction that only binds us in our ignorance if our thinking remains entangled within them. However the climate crisis may shape our future migration patterns, we must be adaptable to all possible outcomes, developing both people and land-centred solutions that support those of us who want to move, and those who will choose to stay.

Beyond this, there is much deeper work to be done on how we rebuild communities and solidarities beyond borders, reconfiguring our relationships and cultivating much more honest and meaningful ways of valuing and treating one another. So that global movement can no longer be informed by racist hierarches, exploitation and harmful systems of oppression.

PART THREE

Land

This part of the book brings us towards the foundation of our struggles: land. The driving force behind many wars and much bloodshed in our world, the ultimate pursuit of the resources fertile land can offer. 'Forty acres and a mule' was the false promise US President Abraham Lincoln made to formerly enslaved African Americans. A clear indication that access to land is the way to some kind of freedom – sustenance untethered to the punitive whims of capitalism. A plot of land, ripe for self-determination.

Today, we feel the struggle of our landlessness as we live on the edges of private land ownership, on top of one another, often in heavily industrialised urban areas. But to live disconnected from the natural world is to forget life's cycle – and, by extension, our own: time to breathe and rest, to be reborn, as spring shows us we must, again and again.

Land may be the root of much of our struggle, but it's also the root of our liberation. In this section, we'll explore our (re)connection to land as a place of refuge and food growing, particularly in times of deep uncertainty and injustice. There's peace to be found in our green spaces, and their invitation calls us back into belonging with the land for our well-being and healing.

6.

Land

It's not often you find yourself trying to figure out how to free a sheep that's got its head stuck in a fence. Learning how to free animals isn't the most obvious life skill for a city person, and there's no second-guessing just how to do it without hurting either yourself or the sheep in question. Yet, one springtime, during our first trip to the Lake District, this was a dilemma that my friend and I were faced with. The sheep, who could perhaps wisely sense our incompetence, panicked, bucking its legs with every step we took towards the fence, until in the end I reasoned that we should report it to someone, somewhere, gesturing aimlessly towards a set of empty fields.

The countryside makes me nervous. Towns and cities have their own codes that are easier to understand, but the country demands a whole other way of being. There are rules, ways of living in the countryside so subtle that many of us don't even realise the ones we're breaking or upholding.

Everyone seems to know each other, which isn't a bad thing – quite the opposite – but it means that as a stranger, and a Black one at that, there's a particular level of scrutiny about who you are. And there is little to no time to decipher whether that interest is one of curiosity or barefaced racism, especially given that the countryside is where racism can be at its most obvious.

Perhaps this is why I find some comfort in the anonymity of a city. While there's obviously racism there, it's more of a textured existence in community with others. The open space of the countryside comes with silence and invisible tripwires.

The discomfort many racialised people feel in the countryside highlights yet another way we continue to feel displaced and sidelined in the imperial core. Whether through the terrors of colonialism and the reordering of borders, the 'keep out' sign of a private estate, or a small country village where no one else looks like you, our lives are determined by real and invisible no-go zones that shape the parameters of how and where we get to live comfortably.

I speak with Josina Calliste, co-founder of Land in Our Names (LION), a London-based collective that looks to reconnect Black and people of colour (BPOC) to land, in both city and countryside. She reminds me of the revolutionaries Thomas Sankara, Malcom X, Kwame Ture, all of whom had significant things to say about the importance of land. For Malcom X, 'Land is the basis of freedom, justice, and equality.'[1] If anything, as he was also keen to stress, this was our first struggle – being dispossessed of our indigenous lands via colonialism. All wars and manner of violence, past and present, are in some shape or form, enacted in pursuit of the possession of land and the wealth it affords.

A clear example of this is in the UK, the motherland of colonial exploits, where 92 per cent of the countryside in England is owned privately. This means the rest of us only have access to 8 per cent of the countryside, otherwise known as 'the commons'.[2] This, of course, isn't widely known, because private landowners reflect a level of wealth and old money that's rarely spoken about. Case in point – the Royal Family, who own large amounts of land in the UK. They make millions by charging public bodies – such as hospitals, local councils, schools and even Dartmoor Prison – for using this land, and yet don't pay any taxes if they don't want to.[3] Which shouldn't be a problem, because everyone loves paying their taxes, don't they?! There's a whole other game of land ownership and wealth at play in this country, one that many of us 'commoners' will never be privy to. Instead, their power operates hidden in plain sight, a constant in the background, in jarring contrast to those of us consistently living at the opposite end of this wealth.

The academic and writer Corinne Fowler's work, *Green Unpleasant Land*, also explores this, highlighting the ways in which ownership of the UK's countryside, green spaces and stately buildings has been bankrolled by wealth from the transatlantic slave trade.[4] Many of our ancestors' blood, labour and exploitation paid for some of the UK's most revered – and whitest – landmarks.[5] And yet, we exist as an afterthought, on the margins of these places, living in more environmentally harmful conditions beyond our control. People in richer areas in the country, for example, have five times more access to green spaces than those in poorer areas,[6] and Black people are 60 per cent less likely to access green spaces.[7] These figures aren't coincidental.

Instead, the majority of us have had to contend with growing amounts of raw sewage being pumped into rivers within our green spaces on a daily basis across the UK and Ireland by wastewater companies.[8] In August 2024, there was a major cyanide spill in a canal in Walsall, near Birmingham.[9] In a time when public bodies of water are becoming more polluted and unsafe for both the public and surrounding wildlife, 97 per cent of rivers in England are inaccessible to the general public due to private ownership.[10] However, who can *own* a river? Who actually *owns* land? The climate crisis is very loudly showing us that nature has its own timeline and way of being that's far beyond human control.

In many Indigenous philosophies, the natural world is to be protected and understood as something to be in relationship with rather than owned; the land nurtures the people as the people nurture the land.[11] Yet, colonial thought normalises ownership and domination, and by legacy creates hierarchies and boundaries in the spaces we live in. Today, this means if you were in the countryside and decided to veer off the path towards a meadow or take a dip in a river, you could be trespassing.[12]

It's no surprise that many of us feel uncomfortable in rural areas. At the smallest level of discomfort, it's noticing that no one around looks like you, and that should something happen, you're in a near-empty field or mountain without the tools of safety you've come to rely on in cities: community support, witnesses, or even a consistent phone signal should a violent encounter unfold. Even the second-guessing as someone walks past, and having to confront the possibility of what *could* happen, creates a level of anxiety that contradicts the relaxing qualities the countryside is supposed to offer. While it's worth remembering that there are, of course, many people who are warm and welcoming in the countryside – open to chatting or, more neutrally, offering a friendly nod in passing – sometimes the worst does happen.

Siyanda Mngaza, at age 20, was racially assaulted and attacked by three people while walking in the Brecon Beacons, Wales. When she and her family reported the crime to the police, she was arrested and given a four-and-a-half-year sentence, while none of her attackers were arrested or charged.[13] Or there's a risk of running into an extreme right-wing group, like the Patriotic Alterative, who held up banners that read 'White Lives Matter' and 'We Will Not Be Replaced' in the Peak District countryside.[14] This fascist party, that's growing in popularity, had been using the area as a place for their team-building retreats, until they were uncovered in 2021.[15] The fact that this is taking place in the countryside, of all places, sends a very clear message about who and what is supposedly being threatened here.

Our cities have had to evolve with us, adapting around our patterns of migration and reflecting the communities that have congregated there over time. However, the countryside can often feel like a preservation of the country's romanticised idea of itself: prominently white, picturesque and 'untouched'. After all, white people make up 97 per cent of the population in rural areas,[16] so in some way this fantasy lies uninterrupted for those who chose to invest in it. Even the increase in Black characters in period dramas, such as *Bridgerton* or the historical drama *Anne Boleyn*, has only started happening relatively recently, reflecting a shift from the racial purity the UK has, until lately, been keen to uphold.

The truth is, Black people have long been a part of the British rural landscape. People have always travelled through the world, passing through or setting up homes, and the UK is no exception. There are records of Black people living in the Sutherland coast of Scotland as early as the 1790s.[17]

Writer and poet Louisa Adjoa Parker also writes about how the West Country has been home to people of colour for centuries – as enslaved people, sailors, teachers, writers, visiting royalty and students. Her book, *Dorset's Hidden Histories*, highlights these records dating back to the eighteenth and nineteenth centuries in the Dorset Parish Registers.

The Second World War also brought in a whole new demographic of African American soldiers, especially around UK port areas, and the families and children born out of this as a result added to this number.[18] Common narratives around migration don't give space for more nuanced realities of Blackness in Britain.

For Josina Calliste, LION is about reimagining the relationship to land we've not been able to have – stable housing to live in, good-quality food, spiritual practice, pleasure and safe places for our children to play in. But this work has come with obvious challenges. Members of LION, who run workshops, skill shares and rural retreats for BPOC, have experienced racism while doing this work, whether through racist encounters on the way to or from a retreat in the countryside, or even on the retreat itself. Josina describes this work as 'trying to find peace in a volatile space'– one that offers the possibility of being in an outdoor community with BPOC and the benefits that it can bring, as well as one marred by conservatism and biting rural racisms.

Perhaps this is why I found myself so nervous with the sheep in the Lake District. We weren't on a set footpath, but in the middle of a field, and there was no one else around. This line between private and public, acceptable and unacceptable, is often so subtle. We knocked on the door of the nearest house. I found myself attempting to be suitably smiley and non-threatening as we explained about the trapped sheep several fields over to the woman who answered.

'Oh, they're Steve's sheep,' she said, as if that explained anything.

Once she told us she'd let Steve know, we rounded off the conversation with some small talk, and she asked us where we were from. When we mentioned we were on holiday, staying in a cottage round the corner, she chuckled.

'I was going to say, you're not from round here, are you?!'

As if I could have forgotten, even for a second.

6.2: Nature therapy – A reclamation

As for many people, the lockdowns at the start of Covid really changed my relationship to nature. It was a time that made me begin to appreciate the refuge green spaces could offer. Finding new parks to walk in became a priority that's now still firmly embedded in my current routine.

The benefits of nature are widely known. Physically, it can help lower our blood pressure and boost our immune system.[1] But it's also meditative, and is shown to help lower levels of depression and anxiety as well as nurture a sense of joy, creativity and increased concentration.[2]

If you'd have told me this pre-2020, I might well have shrugged. I was never really into nature, which is a shame really, because Yorkshire is supposed to be very beautiful – even described as God's own country – with the picturesque nature of its dales and moors. Sheffield, close to where I grew up, is one of the greenest cities in the country. But all this might as well have been on the other side of the world while I was growing up. The Yorkshire I know existed within very rigid parameters, in buses, schools, libraries and church. It was never in the countryside, bar an occasional park excursion or a school trip to the Peak District. It was self-contained, and unspecific to time and space. In many ways, I could have picked up my upbringing and dropped it anywhere.

Evie Muir, author of *Radical Rest: Notes on Burnout, Healing and Hopeful Futures,* is founder of Peaks of Colour, a 'nature for healing' club run by and for people of colour in the Peak District. Evie also grew up near me, in South Yorkshire, in Hexthorpe, Doncaster, a predominately white and working-class area, and similarly, didn't grow up accessing the green spaces of the area. In fact, she calls her connection to nature and the outdoors almost accidental, following ten years of working in the domestic abuse sector and recovery from her own abusive relationship. It was only when she moved to Sheffield in 2018 that connecting to nature became a way of getting out of the house and looking after herself, especially given her proximity to the Peak District. It soon became integral to her recovery.

Founded in July 2021, Evie's keen to stress that Peaks of Colour is less of a fitness club and more of a trauma-informed LGBTQI+ walking group for people of all abilities. Designed to be an informal offering, she describes the mix of people who come, ranging from avid hikers to people who've forgotten their water and who don't want to get their trainers dirty. What started as a meeting for monthly walks is now evolving into yoga and writing in nature workshops as well as forest and sound bathing.

Informality and kinship are important within a bigger legacy of the disconnection to green spaces many Black and people of colour feel. The barriers for Peaks of Colour members connecting to the Peak District, before the group started, ranged from financial obstacles to a lack of accessible transport to explore the area. While there are buses to and from Sheffield's city centre to the Peaks, it's expensive and many people lack the confidence to use them, unsurprisingly because of the racism that often dominates natural spaces, and, by extension, the routes to getting there.

This feeling of high expense and inaccessibility to green spaces resonates broadly. According to a report by the Ramblers, only 1 per cent of UK National Park visitors were Black, Brown or from a minority ethnic group in 2020. An even smaller amount of those interviewed lived within a five-minute walk to a green space, such as a local park, canal or field.[3] So even within the 8 per cent of the common green spaces available to all of us, there are obstacles preventing us from accessing them.

Evie's work highlights the ongoing need for safety in numbers to overcome these boundaries. The steady growth and interest in the group since its beginning shows, according to her, that 'there's no shortage of desire to be in the outdoors, rather a shortage of safe spaces to do that'.

This is what racial trauma and oppression does, it strips life down to its essentials and stops you from being able to live fully within your surroundings. Within many of our families' histories, we were asked to come to the UK, as a part of the Commonwealth, under the pretence that there would be work here, but also a life. Instead of the 'prosperity and employment'[4] the Windrush generation were promised, they were met with low wages and poor housing conditions, not to mention racial hostility and violence upon their arrival.

While many of our countries of origin may lack job or political stability, our quality of life, in terms of access to fresher food, better weather, lifestyle and access to nature, could have been far greater there. The racist conditions here have meant that many of our lives have been squashed into the edges of the country – physically, in the spaces left available to us to live in, but also culturally, in how we are seen and understood.

It's hard not to think about a parallel existence we might have had in our countries of origin without the risk of over-romanticising a reality that would have also had its own complications and oppressions, something many people living in those same countries will attest to. But in the realm of what could have been, there is at least space and possibility, more than what is so often afforded here.

However, there's something in the resurgence of BPOC nature clubs like LION and Peaks of Colour and also Black Girls Hike, Boots and Beards, Muslim Hikers and the birdwatching group Flock Together, to name a few, that feels like a reclamation of space and time in nature, as it should be, available to us all.

6.3: Some justice, some peace

The sobering reality is that many of us will never get justice in our lifetimes. The legal system, by design, is not for us. The word 'justice' within the context of criminal justice is loaded. Historically, Black people were categorised as subhuman and, therefore, all frameworks around *human* rights excluded us in favour of the archetype of humanity: male, cis-gendered, non-disabled and white. If you've made it this far into the book, you get the drift by now. While the law would never explicitly declare its allegiance to this archetype on paper, the proof is in the pudding. And with the consistent criminalisation and high conviction rates of Black people, compared to police officers who can literally kill us without consequence, there's often too little access to justice for the rest of us.

Evie's work within the domestic abuse sector gives her close insight into the failures of the criminal justice system on a routine basis. As a practitioner, with a particular focus on Black queer survivors of domestic abuse, she supports, on average, a caseload of twenty-five people. Part of her job is learning how to support them while knowing that very few will see justice through the current system. In fact, many are likely to come to more harm through navigating an expensive and institutionally racist system that often gaslights their experiences and has low conviction rates.

For Evie, particularly in her personal experiences of domestic abuse, her healing is the thing she's been able to control.

That's where the power lies, in taking your healing into your own hands, as much as is possible. Because any progress tied to the current system we have depends on being believed, on the people who've caused harm being prosecuted and convicted, and on the police and legal system being ethical – things that survivors can't control, either individually or structurally.

So much of the injustice we face is rarely named. Beyond the big things that plague us, that might even make the news, there are the daily microaggressions, the gaslighting and 'debates' about the existence of racism and inflammatory headlines. There are also things that bite quietly, like noticing the racist behaviour of a friend or finding out a musician or artist whose work connects with you on a soul level is a raging bigot. Things that create a consistent feeling of exclusion and discomfort but that are rarely acknowledged under the larger threats of oppression.

It reminds me of the writer Ta-Nehisi Coates reflecting on how our understanding of slavery as a whole overlooks the singular tragedies of enslavement. Behind the generalities of how we understand slavery is an 'enslaved woman whose mind is as active as your own . . . who prefers the way the light falls in one particular spot in the woods . . . who loves her mother in her own complicated kind of way'.[1] Things get lost in the wider story of our injustice. But it all matters, and it all adds up. In amongst the complicated layers of living through racial trauma, our bodies don't know the distinction between the ongoing realties of what has happened to us and our fears of what might – only how it all feels. Chronic stress, and the impact it has on our bodies, can't be divvied up. We live with the cumulative emotional repercussions.

Nature can become fertile ground for a form of healing justice to emerge. Not one that can replace more formal recognition of the injustices we face, but rather one in parallel that offers a path to centre our collective healing. Connecting to nature offers us a way to process some of our trauma, especially given the regulating effects nature can have on our nervous system.

This is not to underplay the need for other forms of therapy, but not everyone can afford them. And for all that talking through trauma is necessary, we can't think – or even talk – our way out of our pain entirely. This is a lesson that's taken me years to understand: giving space for how the body, not just the mind, needs to heal. Sometimes that means being brought back to healing spaces larger than ourselves.

Cities and towns can be busy and loud. And we can't always understand ourselves and our needs when in constant 'response mode' – often a necessity of city life. While the countryside might not be for everyone, we all deserve the space to see who we might be within the quiet and calm settings nature allows. This is why it matters that we protect our local and national green spaces. Beyond the act of resistance in claiming these spaces as much as ours as anyone else's, we deserve the right to get away, even if only to a local park, to tap into different parts of ourselves and our surroundings.

The group Black Men Walking, set up in 2004 in Sheffield,[2] made me wonder what nature could have done for my father, and what it might have taught me to see him in relationship with the outdoors. True to immigrant parent form, he wasn't massively forthcoming about his experiences of racism while working his way through the NHS in a predominately white northern town, but the little snippets of what I've learned are enough. What it was to be the only Black family growing up in that environment was also enough.

While Black Men Walking has now changed its name to Walk4Health Walking group to include more people,[3] I'm struck by the specific kind of trauma that solidifies within that old-school form of West African masculinity. How wordless it can be. And I wonder what nature would have done for my father, in the ways his words couldn't.

For Evie, setting up Peaks of Colour is the first time her work hasn't felt 'oppressive, exhausting and hopeless' but rather a more generative, restorative way of working. We are all surviving something, and nature provides access to healing we all need in one way or another. In her words:

> There's no greater middle finger to the people that have harmed you, than to heal and to be healthy and happy, and experience joy and move on. That's what Peaks of Colour is, it's an avenue to do that . . . a space to turn your phone off, and not get Twitter updates and viral feeds of more Black suffering.

The never-ending loop of suffering our news feeds can keep us in, the daily stressors of surviving late-stage capitalism and our legal systems that don't work for us – all line our day-to-day realities. There's a wisdom and power to the natural world that has seen and survived it all. To have access to nature is to experience something bigger than any issue or problem we're facing. This is a powerful thing to be in relationship with, and an even more powerful tool of repair that's waiting for us.

6.4: *The space to find out*

In 2021, a couple of years before sheepgate, I found myself in another set of large fields in Devon, this time in very different company.

Picture this: a field full of Black people jumping, dancing and whirling under a clear sky – nothing to be afraid of. Our skin and bodies reflecting a moon glow as we danced into the night.

I was loud that evening, singing from a group playlist we'd created ahead of time for a silent disco. There was something about us all playing music separately through our headphones, but together, that removed a self-consciousness I might otherwise have had in a group of people I hadn't met before this point. But it only took a couple of songs, and I was skipping across the grass, jumping up and down until winded, singing until my voice got hoarse to Prince, Solange, Fela, Little Simz, Missy and Whitney. A bold kind of freedom.

This was my experience of The Dream(ing) Field lab retreat, a two-day camping experience for women and femmes of African heritage to reimagine their relationship with land and nature, through rest and collective care, in the wider context of climate breakdown. Run by artists Jennifer Farmer and Zoë Palmer, they describe this work as part 'crucible, carnival, sanctuary, imaginarium'.[1] Central to their work is the understanding that one of the many ways our marginalisation affects us is through our disconnection from the earth and the wisdom it can offer.

As much as I've now come to recognise how important access to nature is in my life, I'm also aware of how alien green spaces can feel. I very rarely camp, unless at a festival, and I couldn't identify a tree or a bird if my life depended on it. But who knows what languages different parts of nature are speaking – to each other, or to us, even? There's a whole realm of knowing and connection I've come to realise is sorely missing in my life, particularly given that the more dominant mindset many of us exist in places humans at the top of the chain, disconnected from the 'lesser' natural world. We could very easily live a life devoid of nature if we wished to, but it gives rise to a stunning kind of ignorance about the world – one that we're likely to feel the consequences of as the climate crisis escalates.

The Dream(ing) Field Lab offers the earth, specifically the British landscape, as a point of reconnection and an opportunity to honour ancestral wisdom and intuitive knowledge that many of us have been denied.

Looking back at the retreat feels like a dream: just over a year from the start of the pandemic, we were all masked on the train there, tested on arrival and no one was weird about it. It's hazy to think about what was, and how those practices of care have been long abandoned.

Wellness was a key intention of the retreat, and we were lovingly taken care of. There were hammocks and foot bathing, fire pits and meditation, guided herb walks in the area, photography and, most importantly, full permission to do absolutely nothing. I thought I'd just be snoozing on a hammock most of the time, which would have been a dream, but it's interesting how good and safe company can be energising, and how much more I did than I thought I would. Even every tent, which had already been set up before we arrived, had flowers in it. It was a retreat of softness, playfulness and communion so rarely offered elsewhere to Black bodies.

Honouring the body is key to the work of The Dream(ing) Field Lab, as is the recognition that in order to nurture our lives, we need time and space to grow.[2] If we are to think fully about our lives and how we might survive the climate crisis, then it has to be a future with some awareness of and relationship with nature – not one of exclusion.

This might look like understanding nature's cues enough to recognise an incoming storm, or signs of high levels of air pollution. It may also mean fighting to keep a local green space or trees on your street, or better yet growing them, which would not only offer shelter from the sun and rain but also purify the air. We can't just wait to be told about this, because it's not a given that official powers will tell us the whole truth; many of us must know for ourselves.

Also, the deep societal shifts and uncertainty we're experiencing in this time mean that we'll have to reconfigure whatever dreams and hopes we have for the shape of our lives. This is a time to be intentional and reflective, lest the chatter of the world shapes it for you. Who do you want to be in all this mess we're living in? What might set your life apart?

As much as this will call for new ideas and convictions, it may also be about going back to ancestral ways of being that have been lost or diluted in our history. These are the answers that live within us, that we will only more fully know if we learn to access the inner worlds that have been waiting for us. This calls for a kind of stillness that for many of us can be hard to access.

Support with this can take form in many ways: reading, meditation and journalling, being in community with like-minded folk and maintaining a spiritual or religious practice. But I think nature is also a resource we can lean on, of any kind, whether large, like woodlands and green spaces for those who can access them, or small, like a garden, a local park, body of water or houseplant. I think it's about using what resources are available to us and allowing ourselves to be nourished by them, so that when we face the world and all its many changes, we can come from a different, more resourced perspective – one that can only be determined by us.

7.

Growing

7.1: The cost of food

Nothing tells me I'm more in my Aunty Era than the way I mutter to myself about the prices of food – especially in Black and South Asian food shops. Growing up in Rotherham, these shops were in such short supply we had to go to Sheffield. They were a lifeline, with hair products, wigs and extensions in one aisle, and stockfish, chin chin and garri in another. Not to mention the mini market of fresh produce – okra, yams and scotch bonnets – on display outside.

These are the shops of the diaspora, the foods of several cultures congregated under one roof. They're still places I consider a lifeline, though as an adult, I'm grateful to have nearly always lived in areas where these shops are more accessible. It's even become a jokey, informal way of assessing where to live: *How long would it take me to buy plantain from here?* A plantain index, if you will – which currently stands at an easy six minutes. Justice, perhaps, for the scarcity of those early years. Though now into adulthood, I find myself tutting aloud at the rising prices of all these foods, inspecting them to microscopic detail and weighing their waning quality in the palm of my hand as I shake my head in disapproval. Hand me a good perfume and a mumu, and I've arrived.

Unless you're wealthy, rising food prices are affecting us all. Prices have gone up on average by 30 per cent since 2019 due to a mix of environmental and political factors.[1] Droughts and heatwaves are affecting farmers' ability to produce good-quality harvests, which has had a knock-on effect on items such as olive oil and sugar. Fruit and vegetables are also in shorter supply.[2] As a consequence, retailers are paying more, which has translated into them charging us higher prices.

The war in Ukraine has also disrupted significant global supply chains. Before the war, Ukraine was a significant world supplier of grains, maize, barley and sunflower seeds, and so their reduced ability to export these products has also hiked up prices, affecting poorer countries in particular.[3] All this is happening while the products we're buying are also getting smaller in size, in what is otherwise called 'shrinkflation'.[4] In short, we're paying more and getting less – as if you didn't already know.

But this doesn't tell the whole story because companies are taking advantage of these global disruptions to supply chains and hiking their prices for higher profit margins. This is a move that can be more easily done, especially in markets where there are only a few companies dominating and establishing the sector.[5] Tesco, for example, is set to make £2.9 billion in profit for 2024–25, up from the previous year where they made £2.8 billion.[6] Simply put, wherever there is crisis, capitalism is going to capitalise.

Climate change also risks the quality of the food available to us, which stands to impact 80 per cent of the global population, in particular, communities in Sub-Saharan Africa as well as South and Southeast Asia.[7] The threat of food insecurity is also increasingly present because we lack a solid and sustainable agricultural foundation. As it stands, industrial farming practices overshadow more sustainable traditional farming methods, which rely on Indigenous knowledge of local ecosystems and an understanding of the ways in which different crops can be produced. Industrial agriculture, as renowned environmental activist and thinker Vandana Shiva writes, 'is not a knowledge system based on the understanding of ecological processes within an agroecosystem;* rather, it is a collection of violent tools'.[8]

When entire forests and ecosystems were destroyed across the colonial world, it also gave rise to the popularity of monocultures: the mass production of one type of crop, such as tobacco, coffee, rubber, wheat or sugarcane.[9] This farming approach is environmentally devastating, as its methods are water-intensive and effectively destroy the soil.[10]

Shiva, in her book *Who Really Feeds the World?*, also adds that underlying the industrial food system and the prevalence of monocultures is an imperial mindset of competition – where only one crop at a time is favoured and prioritised, and any other crops are viewed as threats and therefore destroyed.[11] I find the synchronicities here staggering: the same attitudes that underpin racism and white supremacy are also found within colonial agricultural practices. It highlights just how deep ideas of divide and conquer run within the imperialist

---

* An ecosystem on a piece of land.

psyche – what was done to the people was also done to the crops and land.

Nature thrives in balance. In order to sustain nature and eco-system health, approaches like permaculture and mixed farming methods allow several crops to coexist well together. These methods could enable better soil quality, save water and increase the biodiversity of insects, plants and animals in the surrounding area.[12] This way of growing has to see a plant for more than what it can produce, but also what it can offer to the ecosystem – how it can support the other plants it's growing with, the soil and what insects or wildlife it may attract. It's a piece in a wider puzzle.

Monocultures flatten this wisdom, depleting the soil because there is nothing to feed back into it. It's also a form of agri-culture that's more likely to need pesticides to help grow and protect the crops, as single crops are more susceptible to pests and diseases.[13] The prevalence of monocultures has even led to the extinction of species that can no longer survive from the lack of biodiversity of the area.[14]

So often when we think of colonisation, we think of the loss of people, their progress and their culture. But it's also the loss of a people's nature, their right to grow their own food and the security that this should offer. Destroying people's crops was a tried and tested method of ensuring 'African sub-mission'.[15] But colonialists were doing more than this – they were also messing with the natural order of things: burning down entire ecosystems and rerouting them through indus-trial agricultural practices, something that Vandana Shiva calls a 'world view of arrogance'.[16] This legacy can't be separated from the climate crisis today, given these same countries are most vulnerable to the crisis.

Not only has this historical legacy left its mark, but ecosystems across the world are still being exploited, largely by the same means. Monocultural production methods are still used in 80 per cent of the world,[17] and their environmental impacts are devastating. It's also set the tone for who is fed and who is not, and why poorer countries across the world are suffering by design. Countries across the African continent are rich in resources, but foreign ownership of land, enabled by corrupt political elites, through trade deals and land grabbing[18] means much of this food is exported across the world by big corporations. Meanwhile, local farmers, who harvest these crops, are paid a low wage.

This means that despite all the foods these countries produce, many people that live there suffer from hunger and poverty, creating a two-tier food system. In fact, many of these countries have to rely on imports for their food. Ghana imports tomatoes from Italy and other produce, such as rice, poultry and soybeans, from other countries.[19] Such imports price out independent farmers who can't compete with the lower prices of international trade. Some farmers have had to leave the country to find work elsewhere.[20]

Senegal also imports 70 per cent of its food, including rice, sugar and fresh vegetables,[21] and the droughts in East Africa have led to more hunger and poverty for the countries across the region, which makes these countries even more reliant on exports. Kenya, for example, relies on imported wheat, the price of which is at an all-time high due to the war in Ukraine.[22]

Such deep international and imperial involvement creates overly complicated and disempowering circumstances that disrupt local cultures and Indigenous ways of eating.

This is also the work of globalisation and capitalism, with the global market offering much of the same food choices worldwide, making our diets overlap in more ways than they ever used to. The foods we want particularly for those of us living in the imperial core, influenced by food and health trends, create high demand, which also wreaks environmental devastation.

Chile has been in a drought for over a decade because of the high demand for avocados. Avocado trees require large amounts of water – this has completely dried up the La Ligua River in Petorca province, where many avocado plantations have been bought up by private investors. It's shocking to see a bird's-eye view of this region. The avocado trees are lush green, while the river nearby is ashy and bone-dry.

Locals now have to contend with severe water scarcity, getting sick from drinking the contaminated water they can find and having to choose between cooking or washing.[23] Local farmers who used to grow not only avocadoes but many crops, such as beans, corn and potatoes, have been pushed out by investors who have been able to buy the land at low cost, forcing many farmers and locals to leave the area to find other means of survival.[24]

This is just one example of how high global demand has a colossal impact on land and people, often far removed from the supermarkets we buy their produce from.

To live within the imperial core is to eat food without seasons or origin. We go to the shops, and most of the food we want is there. But not all food is supposed to be produced all year round – or look as perfect and uniform as most supermarket shelves would have us believe. Great efforts are made, often via unsustainable means, to provide us with the convenience many of us have grown used to.

Knowing where our food comes from is the first step to understanding the impact the climate crisis is likely to have on our food. But beyond any impending threats, how can we act in ways that fully recognise the exploitation many growing communities all over the world suffer at our expense? And how could this influence our choices in the way we buy our food – for the better, and for the long term?

7.2: *What can grow*

In Black cultures, we often associate ourselves with royalty – not in a bloody empire kind of way, but more as a way of reclaiming a dignity and regality that is regularly taken from us. I'll take being called an African Queen as much as anyone, but the reality is that many of our ancestors came from land. The agricultural terrain of the African continent and the Caribbean would have been far bigger and more expansive for our descendants, and so farming, hunting and gathering, foraging, plant medicine and tending to the land would have been common gifts and areas of expertise within our lineage.

For many of us, our disconnection from land and growing is symptomatic of the violent rupture many of our families have suffered, and of how, consequently, the lives we now live demand other skills of us. But for Valerie Goode, founding director of the Ital community garden Coco Collective, this knowledge never leaves us. In her words:

> We, especially descendants of African people, need to walk as if we do know [how to grow]. It's within our genes . . . like riding a bike, you never forget. But you just need to be put into that environment where you're able to realise what you already know.

Coco Collective is a community garden space in Lewisham, London, which started on Juneteenth in 2021. It's an Afro-diaspora-led organisation that aims to provide safe spaces to connect, learn, heal and grow through culturally appropriate community events, volunteering and food-growing workshops. Their vision is one of rematriation, which means a return to spiritual values, as well as cultivating ways to honour the earth through balance and repair.

Valerie, who in a previous life was an ethical fashion designer, has had to learn much of this work on the job, which perhaps explains her conviction that we too can learn how to grow under the right circumstances. Her expertise shines through as she talks to me about the garden – which harvests are doing well this year and which have been affected by irregular weather patterns. I'm very aware that it's a bright world away from my own.

Growing up, gardening was a chore – something that my family left until our garden became too unruly compared to our neighbours' neatly coiffured lawns. My memories of being at one with nature were hot summer days hacking away at overgrown plants and ripping weeds from the root. And while there was some satisfaction in seeing the garden afterwards, I couldn't help but see it all as a thankless task, particularly given the inevitable round of culling that would take place the next year.

Since then, into adulthood, my connection to growing has been sparse. Most of the places I've lived in haven't had a garden, which isn't uncommon – statistically Black people are four times less likely to have access to a balcony or garden space than white people.[1] This is precisely due to the areas and types of houses we can afford to live in. Unless you're particularly intentional about living in a place with a garden, it's usually an add-on or 'nice to have', but rarely prioritised above good flatmates and a responsive landlord, affordable rent or a safe enough area. Little did I know at the time that I was fortunate to have a garden to complain about in the first place.

I'm struck by my – and I'm sure other Black people's – alienation when it comes to growing. It's an uncomfortable contrast when you think of how important growing and seeds have been in Black history. I can't help but think of the people who may well have risked their lives to save seeds as entire ecosystems were destroyed by colonial violence – trusting they could plant them elsewhere – and of the grief at what they weren't able to save.

Ahead of the Middle Passage, African women would braid rice grains, millet and seeds into their children's hair, as well as their own, before they were separated, so that their children could eat.[2] It's this expertise and the ability of these women to grow crops on foreign land that is said to have been one of the major ways rice crops and production came to be across the Americas.[3] They were future-gazers, knowing what they needed to survive even in face of the terrible unknown.

One of the priorities of Thomas Sankara, the late leader of Burkina Faso, was to establish the country's right to have access to grow, produce and distribute culturally appropriate foods through sustainable growing methods – otherwise known as food sovereignty.[4] He'd even started to build several national agricultural programmes defiantly independent of colonial control.[5]

This is a legacy the current leader of Burkina Faso, Captain Ibrahim Traoré, is also prioritising. He has increased local production of tomato, millet and rice, has widely distributed farming equipment to support rural communities and made access to seeds more widely available.[6] There is inherent power and agency in growing your own food because, in Sankara's words, 'He who feeds you controls you.'[7]

Nowhere is the intention to take ownership over the food we grow and eat more apparent than in Coco Collective. Their approach to growing is experimental, particularly when it comes to seeing what different foods can grow in the garden. Cucumbers are an easy grow, and flourish in the space as do potatoes, carrots and lettuce. Valerie's mother, who also regularly volunteers, has started trying to grow both a mango and an avocado plant. They are growing them in a polytunnel, which creates the warmer temperatures that produce of this nature needs. They also grow callaloo, dasheen, yacon, watermelon and sweet potato.

Once they build a pond, they're likely to explore growing more root veg, such as yams, cassava and eddo, which tend to need more water. They also want to expand into growing the bush herbs you'd find in Jamaica. It's an experiment, and they are taking notes on what works and what doesn't, course correcting as they go. It's intentional, and a way of developing a connection to the earth and understanding how to work with it. Not to mention how it all sounds delicious.

This is a far cry from those who live in food deserts, said to affect over a million people in the UK and around 24 million Americans. These are areas where it's very difficult for people to obtain nutritious and affordable food. Access to super-markets is limited, with those that are available further away, smaller and more expensive – which make it especially hard for the elderly, poor and disabled people.[8] Highly processed foods last longer and fast food is cheaper, making it much more accessible – and for many, a lifeline. But this system strips away agency, leaving people with what is left to eat, rather than the freedom to make more proactive choices.[9]

In the US, white neighbourhoods have four times as many supermarkets as Black neighbourhoods. This lack of choice can't be separated from the unequal health outcomes and dis-proportionate rates of disabilities Black Americans face when it comes to diabetes, heart disease or other illnesses that are associated with a nutrient-poor diet.[10]

Food deserts – or, as food justice activist Karen Washington argues, 'food apartheid'[11] – is systemic by design, a series of neglectful choices to deprioritise the areas where poorer people live. It creates conditions that force people to make dis-empowering decisions in order to survive. It would be unheard of in the richest parts of the place you live in.

For Valerie, Coco Collective offers some relief from the cap-italist structures we live under, which determine who can and can't afford food and allows food deserts to exist. Instead, she stresses how the land and soil we live on hold the answers to how we're supposed to live together – 'fruitfully and on pur-pose'. I think to live in the imperial core is to live with a façade of convenience. For those of us who live in well-connected places, food is easily accessible and taken for granted. But I can't help but mourn how it has left many of us disconnected from the very thing many of our ancestors fought tooth and nail for.

Times have changed, precisely because of the decisions made by those who came before us. Many of us are now fortunate enough to never have had to make the life-or-death decision to safeguard seeds as a matter of survival. But as the climate crisis continues to destabilise the global food system and change the availability of our food, how many of us would be ready for that, knowing how it will affect our communities the most?

The climate crisis will force all of us to consider what and how we eat. The globalised food system relies on imports and exports of food worldwide, meaning a good proportion of the food in your cupboards has likely come from another country. A climate disaster, whether one-off or ongoing, on the other side of the world, has the capacity to threaten the availability of the foods we know and rely on. War will also be a factor. Wherever conflict takes place, the most vulnerable will be unable to access food, and experience hunger and famine. Any produce that country might typically export will also be disrupted, causing global knock-on effects.

This means we will need to get used to supermarket shelves looking emptier, and people only taking what they need, in a more staggered way. The food hoarding and clearing the shelves of food at the top of the pandemic were a sobering heads-up on how we might behave amid food shortages – or at least the threat of them. I remember being unable to find basic food items for weeks at a time in those days.

We're going to have to cultivate a much more community-minded approach to food and sharing, rather than the individualistic defaults many of us are operating from now. This will be an ever-growing reality as global conflicts likely intensify in the coming years.

Both our environmental and geopolitical contexts nod to a future where food disruptions and shortages will be more likely, resulting in food that will vary in quality and be much more expensive. I already notice this – some food just doesn't taste the same. Valerie also notices a real difference between foods from the supermarket, including organic, and the quality of the produce from the garden.

Coco Collective practises permaculture, which, as Valerie asserts, is just about 'our relationship with nature, and the relationship different plants have with each other'. This, more than anything else, is an ancestral practice. She gives the example of the Three Sisters – a growing technique developed by Indigenous communities across North and Central America.[12] This is where you grow corn, pumpkins or squashes and beans together, each one playing a part in the other's growth. The beans can climb the corn, instead of a constructed frame that would be needed if grown by themselves. The squashes or pumpkins, Valerie explains, provide ground cover that helps maintain better moisture levels within the soil – meaning less water is needed. This is why this technique is particularly beneficial for areas in the world experiencing water shortages. The pumpkin leaves will also provide nutrients needed for the corn and the beans to grow.

When speaking with Josina from LION, she stresses the need for more of us to find and participate in community spaces and growing projects, where we can grow, eat and sell local food, as a more sustainable way to maintain culturally appropriate diets. One of LION's goals is to own land; in Josina's words:

> Our focus for LION has been the right to access land here, and to live and grow in the country that has taken so much from where we're from. Just because we didn't farm here hundreds of years ago, we were farmed *for* here, and were forced into exploitative relationships

Reconnecting to land and growing on our own terms feels particularly important for the people of the diaspora, for whom many of the foods in our daily diets are from all over the world as well as our countries of origin. The rising temperatures we're experiencing in the UK may well enable conditions for more global foods to grow locally, where they once wouldn't have been able to survive in consistently colder climates.

Crucial to LION's work is the understanding that there's an abundance of land – it's just not being distributed equally. Josina gives the example of the use of golf courses, which take up around 700,000 acres of land in the UK,[13] and questions what our diets could look like if we were able to farm on them instead. More than anything, Josina adds, there must be a greater collective attempt to buy and own land, making local councils accountable for how they sell off land to planners and developers for new builds rather than investing it back into the community.

Coco Collective is a majority shareholder in a council-owned space. Valerie and the collective have a good relationship with the council, but there are also challenges in having to navigate its bureaucracy – often resulting in slow changes. For example, they've been waiting two years for an outdoor shelter, so that vulnerable members of the community can have protection from the rain or the sun on hot days.

As empowering a space as the gardens are, being council-owned does mean existing within its parameters when it comes to any changes they may wish to make (or not). First priority for any land the council sells should be given to the communities already living there. Given the vast amount of land some councils own, how could these spaces be better used for community food growing?

Across the globe, Soul Fire Farm is a project that has been exploring this very question. Located in Grafton, New York, it was co-founded by Leah Penniman, the farm's manager and author of *Farming While Black: Soul Fire Farm's Practical Guide to Liberation on the Land*. Soul Fire Farm centres the Afro-Indigenous community and prioritises growing methods deeply rooted in ancestral practices.[14] They focus on regenerative and ecologically sound growing methods that nurture both crops and the soil.[15] When Leah and her family moved to Albany, New York, they experienced food apartheid first-hand, having no access to fresh, nutrient-dense foods in their low-income neighbourhood.[16] In an interview with *Vogue*, Leah is keen to stress the violence of food apartheid:

> A Black person in America is more likely to die from lack of access to their ancestral foods than they are from all types of violence. If you look at diabetes, kidney failure and heart disease – those are all inextricably linked to what types of food a person has access to.[17]

Leah, who grew up farming, invested in some local land with her partner, co-founded the not-for-profit farm, and set about regenerating the land. Today, on their 80-acre farm, they grow fruits, medicinal plants, vegetables, produce honey as well as raise pasture-fed livestock.[18] Food is donated to residents of the neighbourhood directly affected by food apartheid, while patrons can pay what they can on a sliding scale or with food assistance vouchers.[19] Beyond growing, they also have educational programmes and week-long farming immersions that foster skill-sharing and help train young people and future farmers.

Soul Fire Farm is yet another example of the way in which growing from a self-directed, community-led space can lead to transformative outcomes – outside the world of the industrial food system. When faced with the limited food options for her and her family, Leah thought about her ancestors and the choices they had made to hide seeds before the Middle Passage. She took it as inspiration to keep the spirit of hope alive in growing food for this generation.[20]

This is a reflection we all could stand to take time for. Even in times of crisis and deep despair, what seeds do we need to hold on to for the future? What must we protect and grow? Beyond words, how can our actions speak of the volumes of love and protection we wish to offer for those coming after us? What seeds will weather the storms of the climate crisis, and those terrible unknowns we don't yet have the language for?

7.3: Herbalism – Community medicine

Plant medicine is another way we can help heal ourselves in direct connection with the land. Herbalism is the process of using different parts of a plant to help heal the source of a medical issue, beyond just treating the symptoms.[1]

For any cynics among us, if it's widely accepted that peppermint can help with bloating and digestion, that lemon and honey can help with a sore throat and that lavender can ease anxiety, then imagine a world of plants, each with a whole array of different medicinal properties, that we have the potential to be in relationship with.

For Cherrelle Douglas, a herbalist who's been engaging with plant medicine for over a decade, herbalism has been an empowering way to take care of herself and others. It started when she attended a couple of talks given by people who were supporting themselves through fibroids and diabetes with herbs and the food they ate. This kickstarted a journey of self-study and workshops before she went on to study clinical herbalism in 2019. Today, she primarily supports patients with menstrual health conditions – fibroids, endometriosis, polycystic ovary syndrome (PCOS) and infertility – as well as mental and emotional health and autoimmune conditions. She also facilitates workshops and has started leading community herb and wellness walks.

From a more personal place, herbalism has enabled her to live in closer alignment with herself and her body as well as

reconnect with her cultural heritage and ancestral healing traditions. This is something that, for her, has 'been nourishing on
a soul level'– a reconnection she also encourages those of us
who are interested to follow:

> If accessible to you, get to know the healing traditions of your
> own lineage(s). Have conversations with the elders in your
> family and community about the ways they used herbs grow-
> ing up. There is so much knowledge here.

She also explains that the majority of the world – 80 per cent,
in fact – still use plant medicine for their primary health care,[2]
which makes the climate destruction of the lands where these
herbs grow particularly devastating. Plant medicine, Cherelle
asserts, 'is what our ancestors knew'.

This isn't, by any means, to suggest that it replaces modern
medicine, but to recognise the alternatives many of us were
ancestrally raised with as well as the systemic shortcomings
of the medical model. It's a system that too often focuses on
symptoms rather than root causes. The gaslighting many of us
have also suffered is often medically racist and negligent, built
on a shaky racist past in which our bodies have been viewed
as non-human. Maternal health care alone varies significantly
according to race, with Black women being 3.7 times more
likely to die during pregnancy or in the first year after than
white women.[3] I've lost count of the times I've felt awful and
have been shooed off because the 'test results have come back
normal'.

Also, the health care system has failed us massively when it
comes to Covid – a stark reminder that this is a system that

isn't neutral, but rather an extension of the state and its ways of operating. In denying the severity of Covid, how good can our health care system truly be for those of us suffering from it? Long Covid sufferers know all too well how many times they've been disbelieved and left to survive this debilitating illness with very little support. I know several friends with long Covid, who've had to become their own doctor, figuring out their illnesses by trial and error of supplements and medications that may help, waiting for their doctors to catch up to them. This is a familiar story for many people who are disabled and chronically ill, who must navigate a system that at times offers them very little. As other pandemics may come for us, it's not a given that the medical model will move at the pace required – particularly given the lack of funding and resources the NHS is already struggling under.

For those of us too scared to see a doctor, such as migrants or refugees with uncertain status, going to the doctor's might not feel plausible, given the knock-on effects of the hostile environment policies. It pays for us to invest in other means of medical support that work in tandem with modern medicine.

The language of the medical model also doesn't have space for the fullness of human suffering, especially from a Black perspective. Beyond physical illness, anger, heartbreak and grief exist in the body, which can hardly be solved by popping a couple of paracetamol. But these are emotions that, if left to grow in the body, can manifest in illness. By extension, for an asylum seeker or refugee living in the UK, what prescription can truly help them process the horrors they've suffered? We need a far more holistic and embodied approach to understanding our healing, given how much of the violence inflicted on us on a daily basis is wordless. Plant medicine is one way of contributing towards this.

The wellness industry is thriving, and the quest for good health has never been more expensive or profitable. But clearly not everyone can afford wellness retreats, the high-tech fitness trackers or an extensive array of supplements or remedies that claim to heal us. It's also a very individualistic way of looking at health. This is not a criticism of the individuals who seek these options out, but rather a reflection of capitalism selling our health back to us within a flawed health care system. There is, of course, a version of herbalism that also fits into this world – where Indigenous practices are co-opted, disconnected from their origin and sold at a premium to those that can afford them. But community herbalism offers opportunities for collective health and healing at a time when so many of us have been let down and left to grapple with our health alone.

For Cherelle:

> While living in a system that is designed to keep us sick, we can utilise herbalism as preventative care. Healing should be a collective experience. Prior to colonialism, traditional medicine was practiced in community.

In 2020, the Black US herbalist and owner of JamHaw Herbals Jamesa Hawthorne sent free, tailored herbal care packages to Black people amid the police brutality and uncertainty of the pandemic. These packages included herbal blends to use in teas or the bath to target stress, as well as thyme to support the lungs and astragalus to lower blood pressure.[4]

Back in the UK, there are also several herbalist initiatives designed to support and prioritise the community. Community Apothecary, based in Waltham Forest, east London, aims to make herbalism accessible through their local gardens, where they treat people with the plants they've grown – in their words from 'seed to salve'.[5] This work is driven by values of mutual health care and skill-sharing, offering workshops and courses in plant medicine, land care and herb knowledge.[6]

Movement in Thyme, a not-for-profit based in Greater Glasgow and Stirling, Scotland, also works with the local community as well as refugee and marginalised groups both locally and internationally.[7] Central to their work is teaching people how to make and use herbs for their health and well-being. They have their own herbal garden and a community apothecary, where products are available for free or by donation. They also run workshops on how to grow sustainably for biodiversity and community resilience.[8]

The Apothecary Network, a project with Zoë Palmer from the Dream(ing) Field Lab as one of its co-creators, is developing an app that offers recommendations of supportive plants or 'plant allies', depending on the access to green space users have around them and the support they might need. The plants, suggested by the app, will be chosen based on their accessibility and healing properties as well as their effectiveness in addressing historical and climate harms suffered disproportionately by the racialised communities.[9]

There's something powerful about having greater access to medicinal herbs in community settings. Key to many of these spaces is the desire to teach people the medicinal properties of herbs, many of which can be grown locally. This is why access to green spaces is important – whether via an individual garden, community space or allotment.

At the height of colonial violence, when land was destroyed, not only were crops taken but medicinal herbs were too, along with the right to have a much more interconnected relationship with the natural world and our health within it. This, as Cherelle asserts, is the path to self-reliance, and a way we can reclaim the gifts and traditions we have within our lineages. In a system that harms us, community herbalism is an act of resistance.

Those of us short on time or community spaces can also begin to build a collection of herbs to keep at home. These can be bought in bulk and shared between people to help reduce costs. At the start of the pandemic, I gave out little care packages of elderberries, for their antiviral properties, as well as mullein leaf to treat the lungs. The more we honour the intelligence of plants and grow pockets of knowledge about the plant medicine that we need – both as individuals and communities – the more we become our own library, passing knowledge back and forth and growing it further. It's also a way of healing and maintaining connection via community, a way of seeing people and their struggles and tending to them – resources we're going to need in the future.

Cherelle's herbal first-aid kit

In Cherelle's words:

This is going to look different for everyone as herbal medicine is not one size fits all. I would say start where you are. There is some powerful – and ancestral – medicine in your kitchen.

Ginger

Warming and stimulating, ginger supports circulation throughout the body, improves sluggish digestion, relieves nausea and a hot tea helps to ease a cold. Ginger is also great for period pain and is anti-inflammatory for joint and muscle pain.

Garlic

Nature's antibiotic! Garlic is potent medicine, particularly for infections and for the heart. It helps to lower blood pressure and harmful cholesterol.

Add chopped garlic right at the end of cooking so the medicinal properties are retained.

Thyme

Growing up in the Caribbean meant picking lots of thyme sprigs out of your food. Thyme is warming to the digestive system, helps with the absorption of nutrients and is antimicrobial, meaning it helps to fight infection (hence cooking with lots of it). It's also beautiful medicine for the lungs, relieving coughs and shifting mucus. Thyme's antiseptic properties also support [the treatment of] urinary tract infections.

Cayenne

Another warming and stimulating spice that supports circulation and the digestive system.

When used topically, cayenne is an analgesic, meaning it relieves pain. Combine with ginger and infuse into oil for a joint and muscle rub.

Cinnamon

This sweet spice is helpful for insulin resistance and diabetes. Cinnamon helps the cells take in insulin more efficiently, therefore lowering blood sugar levels. It's another spice that also supports digestion, and clears mucus in colds, coughs, and chest infections.

Remember: *safety first with herbal medicine. Do rigorous research and consult a herbalist where necessary. Be extra cautious if pregnant, and on medication.*

7.4: *The returning generation*

I strongly believe that the land wants us back, the land yearns for us and is welcoming us back

Danielle Peláez[1]

I can't help but think of the cyclical nature of our natural world, from our days and seasons to the earth's orbit. Everything is predicated on elements working together in balance, and we, by extension of our human nature, need it too.

For a very long time, we've been operating out of balance, with our rhythms and ways of being disrupted by colonial violence and capitalism. These forces have pushed us to operate in numbness, devoid of real connection to each other and our natural surroundings.

In Leah Penniman's book *Farming While Black*, she speaks about this generation of Black people as 'the returning generation',[2] seeking the land as our ancestors once did and returning to the root to grow something new. This turn does mark a sea change in the UK too, not just with LION, Coco Collective and other growing initiatives, but also in the general rise in popularity of BPOC walking and nature groups.

However, I think it's worth saying that we don't all need to become farmers and learn to grow food. For many, this would be an accessibility issue, as there's no getting around the fact that growing can be hard physical work and not all of us have the gift of the green thumb, as my struggling house plants will attest. But I do think more of us need to be willing to learn, even if just the basics – to explore growing something on a windowsill. And it should definitely be a requirement for young people growing up in a world riddled with climate catastrophe.

Most importantly, we will certainly need to make ourselves more known to local growing initiatives and community gardens in our area, volunteering, supporting and uplifting them in any way we can. This was a major issue for Coco Collective – the need for consistent volunteering.

This is a moment that calls on us to reflect on how we place value on our food and the expertise of the people growing it. For too long, many of us have supported supermarkets* because of their convenience and their capacity to offer produce at low prices – at least until recently. But growing food demands expertise and an understanding of the agricultural world that has long been taken for granted. Those historically low prices have come at the expense of the farmers, both in the UK and beyond, who have been underpaid as a result.

As food prices continue to rise, it gives us the opportunity to invest in systems beyond the supermarket, supporting, where possible, local and national initiatives that pay and treat their staff fairly. Stepping into this also means honouring our seasonal cycles and getting used to products being in and out of season. There is a cyclical pattern for everything, and what

---

* While supermarkets may be the most accessible means of buying food for disabled people or those short on time, here I'm speaking to those of us who might be able to make alternative choices.

disappears returns in time. Lessons we can all stand to learn from the planet's food cycles – and beyond.

Beyond the fear and uncertainty that the climate crisis presents us with lies an opportunity to return to much more human-centred ways of living. This raises the question: what are the life-affirming skills we've been deprived of knowing? And what are we willing to learn? Growing up, I hated gardening because it was less about care and more about fighting against nature – fighting against overgrown weeds and plants I didn't understand. But now, I'm much more open to the possibility of learning how to grow and tend to the land in ways that are generative – ways that require a patience and a deep knowing of the environment.

Our return to nature also opens a possibility for healing and cultivating other ways of seeing the world. Crucial to the work on Soul Fire Farm is acknowledging the disconnection many of us have suffered and understanding that the very trauma inflicted on us via the land can also be a source of healing.[3]

Part of this work includes working with people who've never touched soil before or who've never picked a berry.[4] Beyond the necessity for us to invest in land and support more local growing is an opportunity for stillness and deep healing, which is also a must for our survival. Touching the grass barefoot, sitting under a tree, feeling the wind on your back – for many of us, our reconnection starts here, in small, simple ways.

Valerie Goode also describes the gardens at Coco Collective as a healing space that can offer clarity and peace. In these times, when stress can kill, she reiterates how we need to call on nature as a means to create ways of destressing as a purposeful action.

We don't have to explain ourselves to the land; all it asks is that we merely exist within in it – to recognise ourselves as part of a bigger ecosystem. This is a kind of belonging many of us have rarely been given in any other way, and a gift that nature offers to us indiscriminately. I think that's a powerful reunion – to know that beyond any system, a greater softness awaits, and a wordless understanding that we belong here.

PART FOUR

How We Survive

*We are each other's
business:
we are each other's
magnitude and bond*

Gwendolyn Brooks[1]

Our bodies and minds, much like the planet, are being plundered of their resources, and we are suffering a violent unravelling – coming apart from ourselves and each other. This part of the book locates us more deeply as part of the ecosystem in need of change, as we explore the physical and mental impact of the habits and behaviours we've developed to survive under late-stage capitalism.

Inherent in any attempt to overcome this, as this section calls for, is risk – in becoming safer for one another, making deeper connections within community, finding ways to live slower, more meaningful lives and ultimately telling the truth of who we are, both the good and the bad.

These are terrifying yet brave acts, worthy of interrogation to bring us back to wholeness – risks worth taking.

8.

Community

8.1: *The myth of the monster*

What motivates the human tendency to construct Others? . . .
Why does the presence of Others make us so afraid?

Toni Morrison[1]

For every deepening crisis, we risk becoming more hostile to
one another.

My friend Farha cycled the world alone from London to
South Africa between 2018 until the pandemic in 2020. I, like
many others, was in awe of her, obviously. Not only because
of the physical demands of doing something like this but also
because, as a working-class woman of colour, this was an
adventure very much of her own determination. I was fortunate enough to see her in London at the start of her journey,
and then later in Dakar, Senegal, in early 2019. I was staying
there for a month and she had stopped for six months to rest
and earn some money.

During the time we spent together, she reflected on her travels
so far and what it was to cycle alone as a Brown woman. While

this was a solitary adventure, her experiences relied on other people – the lorry drivers who gave her bags of oranges en route, locals and fellow travellers who advised her on where to stay and which routes to take. It relied upon her pushing past the fear of seeing other people as inherently bad and instead recognising the good in them. What was most striking was the fear that people would express of the neighbouring countries she was travelling towards. *They* were the problem; *they* were the ones to be feared – even though their neighbours had said the exact same thing about them. What is it about creating the myth of the monster in others?

Injustice thrives on this idea: that the people around us are to be feared. This is how dehumanisation can happen – when we deny people the inherent complexities that each one of us possesses. It makes it easier for us to overlook people and their pain and the injustices that happen to them.

Isolation is the breeding ground for these ideas to take root – because the more we become entrenched in our own worlds, shaped by the biases of our upbringing and political leanings, the more it distorts our everyday reality. And so, it should come as no surprise that, in a period marred by endemic levels of loneliness, we are increasingly susceptible to lies – about ourselves and each other.

Nothing can replace knowing our neighbours or having rich experiences with different types of people in our communities. The problem is that without making a concentrated effort to do that, it's easy for our worlds to remain small – as curated as our newsfeeds, never having overgeneralised claims made about people disproved. Until our assumptions become as good as fact.

Oppressive states want us to remain siloed, for us to consider our problems as unique and singular enough that the blame can be placed at the feet of others. Rather than realising our struggles are deeply interconnected and largely due to ongoing government failures and corporate greed, we're pitted against each other.

Immigrants and the trans community are the perpetual scapegoats du jour, and the targets of anti-immigration and transphobic rhetoric and scaremongering. Islamophobia features regularly, as does anti-Blackness. These divisions are all manufactured; a government-made roster of bigotry (and the corporate powers that influence them), woven into loaded headlines and sensationalised narratives. *We* are the problem, *we* are the ones to be feared. But decades of anti-immigration hate speech and racism have never ultimately improved the conditions of those desperate to blame us – and never will.

It's easier to see the myth of this, especially when we are the ones continually cast as the 'monster'. But this can also overshadow the ways we also judge and misunderstand others. I think that because so much of the lives of Black people in the imperial core is spent having to calculate and intuit who to trust, there are barriers to forming connections with others that we don't even realise. But to stay in that space denies us of our growth and our humanity.

Beyond the healthy, life-saving amount of mistrust of the people and institutions who routinely fail us, what claims do we make about others without truly knowing them? Who makes you feel uncomfortable and why? Whether it's someone of a different economic background, gender expression, sexuality, religion or physical or mental condition, how are we living in our biases and reinforcing disconnection with those around us?

This doesn't have to be out-and-out hatred of other groups, as so often is the weapon of choice in extremist thinking. It can also manifest as apathy or failure to consider others at all. Particularly within the umbrella of Blackness, what threats do we feel towards other Black people, living in ways we were told were dangerous? What myths do you need to overcome?

For those of us still Covid-aware, our communities have shrunk drastically over the years. Here something the opposite of monster-making is happening: we're calling on people to come back to their humanity, to make life-affirming choices that see us all as people worth caring about. Some of us continue to seek community with others who no longer practice any Covid precautions, at great risk and expense to our health, while for many more, being in isolation is all the safety we have. Both are forms of loneliness and a lack of authentic community.

We are perhaps among the most keenly aware that while retreating to our respective silos can provide some safety and self-protection, in reality, it leaves us very vulnerable, particularly under the shadow of oppressive state violence and growing levels of fascism.

In a time marred by such divisiveness, the most powerful thing we can do is look for what we have in common, seeking to understand the complexities that live within every person, and recognise how our struggles intersect. There are also other worlds to discover in others, many that might surprise you. Because more often than not, you'll find something of yourself in those worlds too.

The moment we accept the difference we see in other people as inherently wrong is the moment we open the possibility for our lives to also be made wrong, should our identities, in time, be reconfigured as threatening by state powers. Conflating difference with something deviant is an easy way to manufacture tension, and infighting. But unless a person or group, different to us, is genuinely harmful, living by these myths does nothing but weaken our capacity to collectivise and put on a united front against the real monster – the oppressive powers that actually mean to harm us and lessen the quality of our lives.

8.2: Loneliness – A climate change story

In 2019, I created a series of poems called *Loneliness: A Climate Change Story*, which was exhibited in Somerset House in London for its Earth Day season. The title was, in part, meant to inspire intrigue about two seemingly opposing ideas. But really, I was bringing to light how I see both loneliness and climate change as two deeply interwoven issues, overlapping and, in many ways, feeding into each other.

The climate crisis is a result of generations of ruthless exploitation of land, resources and people, happening at the same time as our lives have become more singular and lonelier than ever. As a result, collective community life is shrinking to the point where the WHO has described our rising levels of loneliness as a public health concern.[1]

This means we are shouldering many of life's difficulties largely alone. Just existing has become so expensive and challenging; salaries aren't stretching as far, many of us are disillusioned, sick and burnt out from the false promises of career trajectories that have never delivered, living through crisis after crisis. We're sick and tired of being sick and tired.

The digital world can offer some kind of antidote – a third space that enables us to find and maintain connections, sometimes with people we wouldn't otherwise meet in person. It also facilitates information sharing from people on the ground across the world, telling us what governments won't. But in the space of a scroll, it can so easily turn sour, with unmoderated trolls and hate speech, doomscrolling disturbing content – all taking hours of our lives at a time.

Over time, the digital realm has become increasingly gentrified by corporate advertisements, also changing the parameters of this world and firmly placing us, yet again, as consumers to be sold to, in a mishmash of voices: influencers selling us clean-girl aesthetics – genocide – celebrity gossip – genocide – five ways to rewire your subconscious – reason #842 why the world is getting worse – come get ready with me! – something else the government has done but won't be held accountable for – quiet luxury – genocide – reasons to be hopeful! Content flitting from one reality to another.

Lost in the depths of your specific algorithm, here is another world which you alone inhabit. And before you know it, the room you're in is dark, and it's the evening, with only the light from your screen illuminating your face . . . or is that just me?

Our aloneness, through the lens of capitalism, makes us the perfect consumers. Far from the community and familial structures of, say, sixty years ago, we're consuming more individually and, in return, sharing less. The days of one TV or phone for a family or community are long gone; we're buying more technology, with many of us owning a combination of gadgets and electronics. We're cooking more meals for one and, therefore, consuming more food items – and the single-use plastics that contain them.

Isolation comes with a 'singles tax', meaning individuals bear the sole brunt of rising food prices, housing costs, inflated bills and subscriptions. That said, some of these changes are the consequences of more women choosing to live their lives independently, which feels like a more empowered change. But these shifts have led to significant changes in the structure of our society and a culture of hyper consumption, where we're buying more products – often lower in quality – at a quicker rate than before.

This has a knock-on effect in regions across the world that supply these products, like the Democratic Republic of Congo, where people are heavily exploited, largely because of the ways in which we demand these technological products. Their rapid shelf life means that, sooner or later, they'll end up in landfill, once we've moved on to the next trend our loneliness tells us we need.

Ghana, for example, has become a dumping ground for the fast fashion industry, which is producing way more clothes than the actual demand. 15 million clothes a week are being sent over, with over half the amount unsellable due to the poor quality of the clothing. These clothes end up either left along Ghana's coastline or burned, which has led to increased air, soil and water pollution.[2]

This is how our buying habits, in the midst of our isolation, further fuel exploitative demand and supply chains. On an emotional level, we often turn to buying things as a distraction, to fill the void that living without close community leaves us with, as capitalism whispers to us through every scripted influencer ad that there is a product available for every feeling.

While more of us than ever before see the illusion of it – and generally just have less disposable income to give in to consumerism – it's still psychologically terrifying to feel so isolated during a turbulent time of crisis and conflict. How many of us are more prone in our vulnerabilities to try and find a sense of protection in things? There's always money to be made from our loneliness.

There's also something about Britain and its island culture that mirrors this sense of isolation – or perhaps even sets the tone we all operate with. To describe Britain as an island can feel at odds with the authority it seeks to have across the world. Yet, the UK *is* an island. In the past decade especially – exemplified by Brexit – the country has spouted the rhetoric of 'taking back control', language that leads us to believe that, as a country, it's more powerful on its own. But in the global context of a climate crisis, and the collaboration it requires, this stance feels misguided on many levels. It's this type of toxic individualism, exhibited on a national and global stage, that feeds into our psyches. We're supposed to make it alone, never conceding to our loneliness and our needs beyond what we can buy.

But loneliness breeds an engineered helplessness – it's impossible to look at the scope of issues ahead of us as a singular person and feel hopeful, which means many of us bury our heads in the sand instead. Much of our lives and our struggles could be made better if shared collectively, but our mindsets have to change.

I think this next chapter of our lives needs to be about intentionally finding trusted people to be in community with and living our lives collectively. I say 'trusted' because we will need to trust each other with our time, resources, vulnerabilities and, in times of illness and disability, our lives. This is a terrifying practice of part discernment and faith that many of us aren't used to. But the reward and the necessity outrun the risk by a mile.

We also need to buy less because, despite what we're told, buying in excess will neither save nor heal us. Instead, where possible, we need to understand the fuller picture and origins of where we buy things from and make more ethical choices. If that means that prices are higher, it could look like pooling our resources to buy fewer things but of better quality, more ethical items that last. Of course, in a deeply capitalist world, with the limited choices many of us have, this might not always be possible. But it's about making those changes where and when we feel able to.

Within our broader communities, this could look like supporting initiatives that value the circular economy – which prioritises the full life cycle of an item through sharing, repurposing and repairing existing products. This could be through finding a local repair café, where you can take in broken items, or joining workshops where you can also learn how to fix things. There are also a growing number of initiatives such as the Library of Things or Share Sheds across the country, where you can rent out a number of household appliances or one-off items, like camping gear or DIY tools, and later return them. My wardrobe has been revolutionised by my local clothes swap and has been a really positive way for me to donate my clothes and pick up new ones in a way that doesn't continue the heavy cycle of fast fashion.

These are things that reflect more sustainable ways of being, but they also situate us back into the communities we live in, in ways that are far more meaningful than yet another package left on our doorsteps – seemingly with no origin.

These are also ways, in times of impending world conflicts and crisis, that depend less on global supply chains and their exploitative ways of operating. After all, given the likelihood that global conflict may affect the availability of products we rely on, we will need to practice building local networks and become more sufficient on a community level.

But more than this, moving beyond hyper-consumer mode offers us the opportunity to interrogate our needs beyond the longing that we often attach to the things we can buy. It can be overwhelming, but the remedy is better found within ourselves and in loving community than it is anywhere else.

8.3: *Moving away from the politics of carelessness*

In a sense the state functions like a parent, telling you how you should think and feel and act

bell hooks[1]

Within modern society, we're told we're supposed to be self-sufficient: to vie for that promotion at work above others, get on the property ladder and prioritise success above all else. It's tied to our survival, and for the majority of us who don't come from money, it's the knowledge of how we make it.

Our romantic relationships also seem to fall into this competitive mould and are tightly tied to heteronormativity. We're supposed to take the right amount of time to find 'the one', as long as it's before 30, when the nuclear family structure, comprising a heterosexual couple and child(ren), becomes the next pitstop of desirability – an extension of the individual unit.[2] It's supposed to be tidy, something so self-contained you could pick it up and place it anywhere if, for example, a job required it.

Operating within a nuclear family structure also reinforces a sense of competition. With children to provide for, many of us, by necessity, have to become more single-minded in how we earn money, and must prioritise the family above all else.

All this is very 'normal'. In fact, a deviation from these norms often feels like failing, like missing a turn somewhere. Writer and author of *How We Show Up* Mia Birdsong talks about how we are judged and judge others for how closely they toe this line.[3] Yet there is something so hollow and contradictory about it all – our seeking the very same conditions that crush many of us.

Society is increasingly set up to encourage us to look inwards. Working forty-plus hours a week, on top of carrying out any family and caring duties, strips most of us of the capacity to care for ourselves and others. So many of us are running on empty.

Plus, the history of austerity measures during fourteen years of the Tory government has completely gutted third spaces: libraries, community and youth centres – places that could offer some form of care and physical connection on a community level.

These spending cuts were built on a history of Tory rhetoric that values the free market and individualism above all else. Margaret Thatcher, who famously claimed, 'There's no such thing as society',[4] pushed the idea that care boils down to the individual person and that the responsibility of government for anything more social or collective is a myth. This was reflected at the height of her rule in the eighties, in the shrinking and privatisation of many collective public services that people rely on. In so many ways, we are still living in Thatcher's legacy, both personally and politically.

Over the past decade, funding for youth services has been cut by 70 per cent in Wales and England, leading to the loss of 750 youth centres.[5] Communal spaces, such as day centres for the elderly, libraries, playing fields and leisure centres, have all dwindled as they were either closed down or sold off.[6] These spaces not only served a collective need, but also gave people the opportunity to simply exist with one another – crucially, without being sold to.

Today, third spaces have largely become privatised entities. Online spaces – particularly social media apps – are paid for with our attention and the ways we are sold to, whether subtly or overtly, through adverts. In the physical realm, third spaces now look like paid members' clubs and groups – which, while undoubtedly beneficial to those who use them, position community as an elite perk we now have to pay for.

There's something incredibly violent about the gutting of our public spaces in how it negatively affects the quality of life for the most vulnerable, who most rely on them. It also reflects how the business case for selling off these public assets is so clearly prioritised as a governing ideology over community and collective well-being.

This is also apparent in how our green spaces are understood. Through the lens of privatisation, green spaces can't be left alone when there are more blocks of flats to build, land to sell off or frack – there must always be a profit involved, even while it continues to splinter our communities in the process. As a consequence, we are being ushered towards individual spaces and, by extension, individualistic thinking: you and yours, and no one else.

But it's these ways of being that we must counteract, even if government policy and behaviour normalise them. The government's response to Covid demonstrates the downfalls of making decisions that affect the collective through an individualistic mindset. The 'you do you, hun' approach we've been pushed into assumes that we don't need one another, that my decision will have no impact on anyone else, and vice versa – the very opposite of how an airborne virus works. If anything, Covid is an awfully effective way of understanding our deep interconnection. Yet instead the burden has fallen heavily on the most vulnerable: the elderly and the young, disabled people and the immunocompromised – the latter categories widening to include more of us as the pandemic silently continues.

For the writer Naomi Klein in her book *Doppelganger*:

> [Covid] was a crisis that could only be met if we chose to truly see one another, even those labouring and living in the shadows. A crisis that could only be addressed with collective action and a willingness to make some individual sacrifices for the greater good.[7]

This is a clear example of how following government norms harms us, given the individualistic, back-to-business ideologies they operate from, that leave so many of us uncared for in the process. In a similar way, if we had a government that was actually responsive to the challenges of the climate crisis at its first inklings, we'd be in a very different position to the dire one we're in now.

There is a heavy realisation that comes with accepting that the institutions we've been taught to rely on can't and won't save us in their current form. But I think often of bell hooks's words about the parental role of the state – how its decisions overshadow how we convene, where we place value and how we support each other.

Much like the young adult developing their own identity away from their parents, we need to build communities of support in our own image, separate from the government's immoral frameworks and cultural norms. Because in following them, we become the physical embodiment of their mistakes and we become agents of their harm – often at the expense of our own standards and values.

What kind of lives are waiting for us on the other side of mainstream societal norms that take us away from our humanity? Instead of waiting for some grand plan when it comes to the climate crisis, what plans, education and awareness-raising can we develop on a community level?

I know that in times of great difficulty, toeing the accepted line of normality can often feel easier, but that's how we end up in isolation, and internalising our struggles. With no real clarity on the care and support the government will offer when it comes to the climate crisis and beyond, how do we bring care back to one another, and prepare ourselves and our communities to show up for one another as a regular practice?

8.4: Knowing each other's names

In the absence of so many support systems, we are it for each other

Toni Morrison[1]

In May 2021, around 200 people surrounded an immigration enforcement van with two Sikh men inside on Kenmure Street, Glasgow.[2]

Immigration raids are an ongoing strategy used by the Home Office to target so-called 'illegal' migrants in the country. While these raids tend to take place in public spaces, such as in places of work, there are a growing number of dawn raids taking place in people's homes in the early hours. The cruelty behind this is to 'catch' people in moments of vulnerability, when they least expect it, and this van raid was no exception. Locals in the area surrounded the van, shouting, 'These are our neighbours, let them go,' while local shops and neighbours passed around food until 5.30 p.m., when the men were released.[3]

Similarly, another raid was stopped in Peckham, London, where protesters blocked a van attempting to arrest and take away a Nigerian man for allegedly 'overstaying his visa'. Protesters encircled the van, sitting down as the police attempted to push through them, and the man was later released on bail.[4]

Community power works. Strikingly, the impact behind anti-raid protests is the snowball effect: someone has to notice the disruption before spreading the word for others to come forward, which turns into unified action. Crucially, for this and other acts of community care to work, we have to know one another.

As an introvert, I feel a certain level of contradiction in thinking about the importance of this – I am far from the person at the neighbourhood street party who shows up with a quiche. But I don't believe we all need to fit into the same mould and operate in the same ways to build connection. I do, however, believe we all need to do something, in any way we can, whether it's through an anti-raid group, local volunteering, or just making yourself known to your neighbours. There's a level, no matter how small, to which you can seek to know and be known in your area, because ultimately, we need to know each other's names.

Community care is at the heart of how we resist injustice and the realities of the number of ongoing crises we're living through. It means showing up for the concerns of the area and your neighbours locally, as well as looking towards the global. Making connections between both also means showing up through boycotts, and supporting strike action and protest. With this comes a faith that should you too need support, your community will be there for you. No one is supposed to do this alone.

The #SayHerName movement aims to bring the names of Black women in the US back into the mainstream eye – women who are all too often killed and forgotten within narratives surrounding police brutality and systemic violence.[5]

Similarly, thousands of people protested following the abduction of Turkish student Rümeysa Öztürk from Tufts University in the United States. She was taken via an unmarked SUV to a detention centre by plainclothes ICE agents after Homeland Security revoked her student visa with no prior warning.[6] Protesters carried banners demanding Rümeysa's release.[7] Knowing is power. Remembering is power. Saying people's names – keeping them in existence – matters. It proves we're here and can't so be easily forgotten.

Violence against the most vulnerable in society only happens so readily because of the bullish belief that our oppression can happen without consequence. The trans and queer community are far more likely to be on the receiving end of violent crimes and state violence, and police brutality is deeply embedded in the lives of Black and Brown communities. Disabled people are also more likely to be victims of sexual assault,[8] within the wider context in which five out of six women who are sexually assaulted don't feel able to report it.[9]

While the necessary fight for systems and governments to recognise our rights must continue, we can't wait for it. If we don't matter to each other, then it becomes far easier for the corrupt powers that be to pick us apart, confident that their violent and suppressive behaviours will be of no consequence if they choose to harm any of us. Forging deep community bonds is one way we firmly send the message that this can't happen.

Recognising our humanity needs to start with us.

8.5: What Black community means

I'm aware that community care feels like a bit of a buzzword at the moment, and it's certainly a topic well explored by our foremothers. For Audre Lorde, 'Without community, there is no Liberation.'[1] bell hooks also writes in detail about the importance of community love and how the small 'privatised' unit of the nuclear family makes us more susceptible to isolation and abuses of power.[2] In more recent thought, adrienne maree brown asks us how we can put our collective minds towards collaboration as a means to survive[3] beyond 'violent competition'.[4]

Black people wouldn't have survived without community. At the heart of the Black Panther Party were community programmes to support the needs of the community, including education and health classes as well as breakfast clubs.[5] They were also reactive and set up a free ambulance service after a fifteen-year-old boy died after being shot and denied medical care by the county ambulance.[6] They established a co-operative housing programme, legal aid[7] and a free plumbing and maintenance scheme to help repair people's homes.[8]

The idea that any one of us could learn a craft, skill or even profession to serve the community directly feels like switching a light back on. These are ways of operating that our communities have long been engaged in but that seem to have become diluted in more recent times.

The 1950s onwards was a time of marked racial discrimination towards Black communities in the UK, including exclusion from secure employment and housing. This led to the birth of alternative ways of saving money and building resources in community. Money co-operative schemes, where groups pool their money and borrow from it at low interest rates,[9] were formed, such as the Hornsey Co-operative Credit Union in 1962. Set up by ten Jamaican members of a Baptist church in the area, they did this in response to being unable to secure loans at the banks, which discriminated against Caribbean people. If they could access loans at all, they were at much higher interest rates and often under the condition they pay higher deposits on their mortgages.[10]

Informal ways of saving also took hold, through schemes like susu or pardner, where a group of people contribute a certain amount of money on a monthly or weekly basis to a collective fund. After an agreed time, each person gets a turn to receive a lump sum from the pot.[11] Not only was this an interest-free way to build savings, it also helped those excluded from main-stream financial services to make bigger purchases in a way that would have been far more difficult to achieve solo.[12]

These are just a few ways Black communities have worked collectively outside the system as a means of survival, but many of us will have similar examples within our families and communities. Whether it's having a cousin or aunty stay when they've just arrived in the country, or babysitting or cooking meals for a sick relative, we've always had ways.

Community, and the extensive web of people we consider extended family, has been the backbone of our survival. This, of course, hasn't disappeared overnight; there are many events, groups and networks that have been formed by and for Black communities to come together, which offer us some much-needed relief.

But I do wonder how many of these moments we're creating for convening still have a proactive element of building together. How many of us would save together, buy homes and raise children together? How many of us are building deep relationships of intimacy and trust that would enable us to think and build together in this way, beyond the conventions of a nuclear family?

Perhaps the closest we have to collective support is crowd-funding, which helps people in standout times of need and emergency. This never fails to show the potential of our collective responsiveness and compassion – but how can we mobilise this not just as a last-resort option but as a way to shift towards creating support systems in each other, as our larger societal support systems continue to fail us?

Since the 1950s, life has become materially better for Black people, and we've made strides in employment and public life that many of our elders could only have dreamed of. But racism hasn't disappeared – it's simply taken on more insidious forms. Our struggles may have evolved, but our need for community has not.

I wonder whether our improved material conditions have led to a complacency and loss of creativity that living on the sharp end of oppression has historically generated. I'm reluctant to romanticise struggle; but it often leaves a person or group with no choice but to lean into kinship and developing alternative ways to survive. I wonder if, in the fight for equality, we've lost sight of better ways to support ourselves and our communities than the mainstream could ever offer.

Queer cultures are a great example of life beyond heteronormativity and individualistic ways of living. The ongoing violent rejection of queer communities within mainstream cultures, while deeply traumatic, has led to other ways of knowing and building love. Creating a chosen family – based on many kinds of love, platonic, romantic and familial – means building on shared realities and values. It reminds us of our agency: we get to choose the fabric of the world we live in, and build with others who see and support us, beyond biology, beyond law, and ultimately beyond our oppression.

9.
Slow Down

The black body was Europe's first unit of energy, so the relationship between exploitation and resources, for those of us of African descent, is a very direct relationship, it's felt in the skin

Professor Lesley Lokko[1]

9.1: There is no separation

The earth is on fire. To be dysregulated is to be watching. To be dysregulated is to be listening to the Earth's regulation. It is to be in sync with the panicked heartbreak of the world; sounding the alarm over and over again

Project LETS[1]

When I say we are the ecosystem in crisis, I am speaking of non-separation – the intricate connection between our bodies, the demands of capitalism and the destruction of the environment. We exhaust and push ourselves under the same hand of capitalism that plunders the world's natural resources – both our capacity and that of the planet are finite and at breaking point.

The idea of non-separation can be understood in both the best and worst of ways. At its best, it's in the interconnected nature of things – how we can only exist on this planet because of each other. Of all the people, circumstances and places, it was two individuals who brought you into the world to live during this time. It's the land, skills and expertise of people who grow the food we eat daily. We are here because of the trees, which not only bear the fruit we eat, but also convert carbon dioxide in the air into the oxygen we breathe. It's in how we are grounded by being in nature, and how it restores us and brings us back to equilibrium.

But at its worst, non-separation, through a capitalist lens, understands human beings as a living extension of the natural resources to be exploited. After all, HR departments in the workplace stand for *human* resources. This is how we've always been understood. No clearer example of this exists than transatlantic slavery, which, besides its many other horrors, was a business – one of pushing Black bodies to the point of indefinite servitude as capital to be plundered, sold and traded. Much like the ongoing legacies that modern slavery and human trafficking uphold, there is always a price at which we can be sold.

The relentless extraction of the planet's resources speaks to both our material and bodily worlds – the exploitation of natural reserves mirrors the extraction of our minds, bodies and spirits, taken to churn the cogs of the capitalist machine, at the expense of our own livelihoods and well-being. Our bodies are routinely sacrificed for capitalism, which, underlined by ongoing racial discrimination, builds on legacies that see our labour and our servitude as our singular purpose.

Capitalism today continues to demand our time, energy and labour in exchange for financial compensation. It offers the promise of freedom through following its rules, yet it barely delivers – at least not for the majority of us. We see this in the countless stories of people giving years of service to an organisation that fires and replaces them in a heartbeat. How many times have you felt just as disposable or disregarded in your work? Especially when you become 'too Black' or 'too sick' for the system to contend with you?

For the public intellectual Noam Chomsky:

When you have a job, you're under total control of the masters of the enterprise . . . The very idea of a wage contract is selling yourself into servitude. These are private governments; they can control everything that you do . . . you have a choice between starving or selling yourself to a tyranny.[2]

And yet, even a job is no longer a guarantee of financial security. In-work poverty is on the rise.[3] Despite working full-time, most of us can't afford the steady rise in rent and bills – never mind the freedom to pursue any bigger goals.

While the system is repeatedly failing us, it works very successfully for a small majority, which is why, for all of its glaring inadequacies, very little changes. Instead, the goalposts move, and we're hit with a number of reasons for our supposed shortcomings. These range from the claims we're just not working hard enough (a classic), to more ridiculous accusations – such as the reason younger generations can't afford a deposit on a house isn't because of the absurd rise in house prices, coupled with salary stagnation, but because they buy too many cappuccinos.

Instead, we're told to grind, to give more to the machine and it will *eventually* come good. But the reality is, for the majority, it never will – the carrot and stick in action.

Slowing down is how we resist. It's how we reclaim our personhood, while contesting the ways the system seeks to actively dehumanise us. The less we tie our self-worth to our jobs – the less we define ourselves by promotions, productivity and professional success – the more possible it becomes to slow down in them. I know this is easier said than done, and that it's not a complete truth. In certain professions – especially the medical health care sector – urgency is crucial. Here, in fact, it's the slowness in responding to our health care needs leading to long wait times that's killing us.

But a shift – for those that can make it – is already happening.

The growing phenomenon of 'quiet quitting' – where employees consciously put in less effort at their jobs, committing to doing 'just enough' and no more – reflects this. More of us are recognising that hard work is rarely acknowledged, and promotions often lead to more responsibility, disproportionate to any rise in salary. Calls for a four-day week and people's preference for remote working recognise, in part, the benefits of slower living.

Working from home during the lockdowns also offered some people more of a work–life balance: slower mornings, being able to pick the kids up from school – all while being as, if not more, productive at work.

Remote work also disrupts the performance of professionalism, of keeping seemingly busy and staying late, *just because everyone else is*, and shifts the focus towards tasks that need to be done – no more and no less.

As much as I'd love for more of us to be able to actually work less, most of us lack the financial privilege to do so. But this is as much of a mindset shift as it is a material one, a commitment to move past the cultural norms where what we do for work is the most central – even most interesting part – of who we are. It's a call to move past sacrificing too much of ourselves because of it.

The resistance that many corporations have shown to their staff working from home, despite evidence that it's a more productive way of working,[4] is economic. Fewer people in the office means those who have bought properties to rent as office space are seeing a lower return on their investment. It also means we spend less on commuting and other outgoings, such as work clothes and food on the go.

But beyond this, I think it also poses a philosophical threat. They want us pushing our real lives to the outskirts of the working week: gym in the early hours, childcare, community care and political action after work and at weekends if we have the energy at all – with the best parts of us, our minds, energy and ideas depleted.

A former boss, at a job where I was earning £17k a year, asked me and my colleagues to keep notebooks by our beds, in case inspiration took hold in the middle of the night. This was a particularly dull job, which made the suggestion even more questionable. But I've also worked myself into the early hours for jobs that have barely batted an eyelid at my contributions. The entitlement that many places of work feel like they have to our to time is astounding. Writer David Cain, in his article 'Your Lifestyle Has Already Been Designed', puts it simply:

> [T]he 8-hour workday is too profitable for big business . . . because it makes for such a purchase-happy public . . . We've been led into a culture that has been engineered to leave us tired, hungry for indulgence, willing to pay a lot for con-venience and entertainment, and most importantly, vaguely dissatisfied with our lives so that we continue wanting things

we don't have. We buy so much because it always seems like something is still missing.[5]

This sentiment is also echoed in the late David Graeber's book *Bullshit Jobs*. Graeber argues that beyond any economic reasons that drive the system to make us work in the ways we do, there is a moral and political understanding that letting us have more autonomy and time over our lives is the beginning of a dangerous idea.[6]

The term 'bullshit jobs' is based on a YouGov poll in which 37 per cent of people thought their jobs were pointless, unnecessary and meaningless. But as office cultures and standards of professionalism dictate, many must go through the motions and play along.[7] In these instances, slowness and decentralising work is entirely possible.

As we collectively wake up to the false promises of much of the working world and all that it takes from us, this growing fatigue and dissatisfaction are being mirrored by the planet at large. Its response to the continual exploitation of its natural resources manifests as extreme heatwaves and unpredictable weather conditions, floods and storms. Our simultaneous rage, discomfort and protest are interconnected, as will be the ways we look to overcome and heal.

But first, we must slow down.

9.2: I am not a machine

I moved to London in my twenties to make something of myself – reluctantly. I say 'reluctantly' because I was born to immigrant parents, whose sacrifices, in many ways, had been made for the success of this very period of my life, and I felt the expectation of this very heavily.

It wasn't necessarily that I didn't want to work or find my own path in life, but I found the idea of being a workhorse – of clocking in and out, with only the weekend to look forward to – particularly unappealing. I'd had glimpses of working life as a student when I worked in hospitality, retail and admin and had found it all exhausting and uninspiring.

Yet still, when I came to London, I became the workhorse I'd feared, living in a box room in Elephant and Castle* that I barely spent any time in, working seven days out of seven, and doing jobs and internships for very little money. I packed breakfast, lunch, dinner and snacks for the day in my backpack, and walked everywhere to save money.

I felt a great deal of shame during periods of unemployment between jobs. I'd bought into the idea that my productivity was tied to my worth – never mind that those jobs did very little for my sense of worth, either. Still, work was the marker

---

* This was during the pre-Pret stages of gentrification of the area, which is now unrecognisable.

of my value, my usefulness – proof that something had come good of the risks my parents had taken.

As time passed, work became more consistent, and the busyness continued. I began working in the NGO sector while studying part-time for a master's, and simultaneously building a career in poetry – all while trying to have something of a social life. Sleep was not restorative in those days; it provided just enough rest for me to carry on giving to everything but myself.

There were times when I caught myself muttering 'I am not a machine, I am not a machine' while writing an essay in the early morning before my nine-to-five, or on my way from work to perform at a poetry gig in the evening. I find it incredible that long before finally catching up with myself – long before realising my exhaustion was unsustainable – there was something in me that knew all along: that my initial reluctance meant something more than laziness. That our bodies and minds, no matter how deeply conditioned to think otherwise, know their right to rest. That we only need to stop long enough to listen.

Repeat with me: *I am not a machine. I am not a machine.*

9.3: The performance of urgency

I believe the powers that be don't want us rested because they know if we rest, eventually we're going to figure out what is really happening and overturn the entire system. Exhaustion keeps us numb, keeps us zombie-like, keeps us on their clock. Overworking, and the trauma of burnout continues to degrade our divinity

Tricia Hersey

For people working zero-hour contracts and heavy shift work, or those who are parents or caregivers, slowing down can seem like an unreachable idea. I know at the busiest points of my life, I couldn't have heard anything about slowing down without thinking it was condescending. But for those of us who can slow down and, even better, decentre the paid work we do, we'd be more available to support the people in our communities who need it most, which would alleviate some of the pressures they have, and free more of us collectively.

In fact, research shows that in order for all of us to have a good standard of living worldwide – meaning equal access to shelter, hygiene, nutritious food, clothing, health care, recreational activity and education – we only need to use 30 per cent of our global resources.[1] Yes, read that again. It's entirely possible for us to slow down, live stable and fulfilled lives, and address environmental harms. It is, in fact, the answer.[2]

Beyond any real responsibilities we have, urgency – particularly within most professional contexts – is often a performance. So often we're praised for our ability to work in 'fast-paced environments' but that pace is rarely set for any other reason than to produce outputs and profits for others. Beyond issues of life and death, most deadlines are made up – only meaningful and consequential because of the expectations we and others place upon them.

While this is probably not going to land with your boss, the truth is these so-called burning deadlines are a fiction and cultures of 'working around the clock' are wholly unnecessary. Not least because those expected to give the most of themselves to the clock are also the most marginalised – the Black, Brown, and working class.

Within these jobs, we're having to navigate code-switching, having to prove ourselves more than others, and within office cultures, being inevitably roped into an unpaid task force around diversity. We're perceived as lazy when we don't stand out, aggressive and overbearing when we do. All this toxicity contributes to stress and illness, as we are repeatedly pushed beyond our limits.

Grind culture also promotes and normalises a state of always being on the go, and will have you working those full-time hours while managing several side hustles. I know that this is precisely to escape the drudgery and microaggressions of working for others, not to mention the necessity of needing to earn enough to survive because we're so often paid less. But the cost is too often our health and our ability to rest. We shouldn't have to work to the point of self-destruction just to exist as human beings. *Any* work that we do should be enough for us to live. This is a failing of the system.

It's also very hard to spend forty-plus hours working in a pressurised working environment and not have it spill over into other aspects of your life. It marks your way of being. Urgency is addictive and rewires the nervous system. If you've ever lived in, worked in or visited London, you'll know it's common for groups of people to run like it's the final of a 100-metre sprint to catch the Tube – despite the next one arriving only a minute later. I, too, have caught myself running in that crowd many times, panting on the Tube, having just made it – despite there being no real urgency to my destination. It's baffling how the chaos catches you somehow.

It's taken me a long time to notice how urgency continues to trigger anxiety in my own life. That for a long time, even on days off, I've moved with a sense of unnecessary urgency, despite having no external commitments. Life admin and small tasks became stressful and something to accomplish with all the fanfare of a military operation.

I've had to proactively regulate myself and let my body know it can rest and slow down, because it's been running on frenzied autopilot, with no distinction between work and time off. It also spills over into what we expect of others – whether it's the expectation of prompt replies or requests from loved ones, or beyond this, quick food deliveries and packages. This constant availability has gradually become the accepted way of being, so normalised that we then enforce these expectations on ourselves and others.

We are living through in-between times, where different ideologies, philosophies, and old and new ways of being are interacting and clashing with one another. We can believe in slowing down all we want, but your boss is still expecting you in the morning. Bills still need to be paid. Children still need raising. The world still demands of us what it does.

Being able to shift to a slower existence – within our energetic means – isn't going to happen overnight, but it does start in our knowing why it's necessary and placing value on it as and when we can. It begins in our shifting our values and debunking the ones we've been told to live by. It's in repeatedly asking ourselves questions like: *Will I truly not get the promotion because I took the full lunch break or left work on time? Or is that the carrot of capitalism dangling over my head?* Which other falsehoods within the grind of capitalism are fooling us? Even for those of us who like our jobs, how do we lessen their power over our lives?

Let's start slow and take our time back where we can. Whether it's in taking an extra five minutes for yourself before getting out of the car and opening the front door, taking an extra minute to be present with yourself and enjoy the water in the shower, going for slower walks in nature, having long phone calls or taking up a new hobby, let's luxuriate in the privilege of our own company and the people that nourish us.

Time is not a reward, but a right. Understanding this means we'll know when to fight for it.

9.4: Slowness for the revolution

If we're really shackled to ideals that are . . . profit-driven, but not centred around humanity . . . nature . . . collective healing and growth, then we're actually building things that are going to fall apart

Latham Thomas[1]

The times are urgent, let us slow down

Bayo Akomolafe[2]

Slowness stands in contradiction to the urgency of our times. Despite the continued expansion of gas and oil companies worldwide,[3] we've been given a terrifyingly small number of years to prevent the 'irreversible damage of climate change',[4] and the graphs and statistics on the crisis get more irate as the years pass. The rise of fascism and the extreme right also continues to escalate amid genocides and global violence.

Calls for change evoke ideas of revolution, all underpinned by urgent levels of movement and action. But the reality is, we're already in revolutionary times. The movements and revolutions that have inspired us throughout history unfolded over long stretches of time.

The Haitian Revolution took place between 1791 and 1804, and the US civil rights movement spanned from 1954 to 1968 – both more than a decade long. The moments that we learn about are the points of escalation, of rapid action, but they aren't representative of the whole era. We don't often learn about the silences, the slow and consistent moments or the times of rest – but they were just as important. They were what also sustained the moments we hear about.

Urgency is a tactic, not a state of being. Without intention, urgency becomes a distorted kind of normality and a punitive expectation that we achieve more than the limits of our ability. We have to be able to stop long enough to know when and how to move with the urgency required for the right moments. For example, the BLM uprisings following the murder of George Floyd in 2020 required a swell of mass action, which prompted both cultural and political shifts in the collective. But it's impossible to sustain that level of intensity consistently, which calls for a strategic understanding of when and how we move.

Too often, running consistently at a fast pace leaves too much and too many behind, and replicates the same harms we already suffer from. This can look like holding events and protests without accessibility measures in place for disabled communities, running talks without due representation of people most affected on the panel or in the audience. Or developing narratives that miss perspectives of the marginalised communities most affected.

Going too fast under the guise of 'not enough time' means we get sloppy and default to harmful ways of being. Of course, we all do this – present company included – and we will continue to make many mistakes because we belong to a world where collective care has been devalued, and our complex needs are always changing.

But if we're serious about undoing the oppressive behaviours within the society we live in – ableism, sexism, classism, racism, transphobia and queerphobia – then we have to accept the time it truly takes, precisely because of just how ingrained these types of oppression are in every facet of our lives. You can't go fast and unlearn these things at the same time – running at this false sense of urgency hasn't solved anything so far. And if urgency worked, wouldn't we have arrived already?

Slowing down means learning how we all move *together*, learning from each other's perspectives and strengths. It means not rushing to know all the answers, and more listening over speaking – particularly if speaking is what you're more used to doing. It means that the calls for change include all of us, not just those with the loudest voices or easiest access to the microphone. Who wouldn't want to join a movement that speaks directly to their struggles, with others who also experience their struggles firsthand?

Oppressive powers also use urgency and slowness for their own gain. They devise ways to make us work harder through incentives like bonuses and benefits or by fostering toxic high-pressure working environments. We're part of the machine to churn. But slowness can be weaponised against us, too. When there's a question of a pay raise or salary negotiation, a need to talk about racism or discrimination in the workplace, or even a proposal to bring more diversity into senior management, watch how the cogs of the machine grind to a halt. Words like *due diligence, under consideration, process* and *risk assessment* appear.

Similar roadblocks have also been used in our social and environmental movements, albeit in softer language like *taking stock, holding space, listening and learning, reflecting.* These aren't inherently bad actions; much of these behaviours are recommended. But for those of us on the other side of this consideration, those, for example, living in climate catastrophe or under racial injustice, whose lives depend on people's capacity to act and change, know only too well who is play-acting social progress and who actually means it.

This kind of slowness – where the 'reflecting' becomes the final destination rather than a passage to tangible change in action and behaviour – can be the literal death of us. In these situations, the words of James Baldwin come to mind:

> You've always told me it takes time, you taken my father's time, my mother's time, my uncle's time, my brothers and my sister's time, my nieces and my nephew's time. How much time do you want for your progress?[5]

Much like urgency, slowing down should also be intentional work, not a ploy to hinder our evolution. Slowness is an opportunity for you to meet the honesty of your needs and those of others. It's a place to reflect on you and your place in the world you live in. Do you like who you are and how you're showing up for yourself and others? It's a reset, a space to dream of a world other than this – so that when the time comes, once again, to move, you'll know how.

9.5: Being safe to one another

What stories are we holding deep inside that are untold and uncovered because we are too exhausted?

Tricia Hersey[1]

For many of us, there are demons that lurk behind our keeping busy. The idea of more stillness means confronting the things we are running from. Trying to slow down when we're dys-regulated, jittery and wired from the pace of life we've been pushed into is why many people find the idea of slowness unbearable – like trying to lie down after coffee.

Even when we manage to find physical rest, we often spend that time stuck to our phones, distracted and disassociated, too tired to move from the sofa. It's a form of physical rest, but our minds are still running, belonging to someone else via addict-ive content on our phones. It does nothing for the soul and can still leave us very unrested.

Living under capitalist and oppressive structures implicates all of us. Under their shadow, we are both harmed and cause varying levels of harm. Those living within the imperial core, in particular, are raised to not think about the consequences of our choices. We buy products with no real knowledge of the laborious process or exploitation required to bring them to us. We're time-poor, meaning we rely on convenience over items from a more ethical and environmentally sound source, which, for many, are unaffordable anyway.

Capitalism is complicated because it requires our harmful participation, giving us very little choice to opt out. Inherent to surviving under capitalism is ignoring its wrongdoings – and by extension, our own. While this may be necessary for capitalism, it makes us very unsafe to one another, especially when trying to build authentic human connection.

The emotional impact of what capitalism does to us – the long hours and low pay, the humiliation of daily microaggressions and racism, the lack of agency – also festers in the body and spirit. This can manifest in physical, verbal or emotional abuse towards those around us. This isn't to excuse personal responsibility, but rather to explain what living within a violent oppressive context can give rise to.

How many of us, as children, knew when our parents and caregivers had had a bad day at work? Knew how and when to keep our heads down, never knowing the details, only that something had happened? The whole family unit and community are affected under the brutality of the society we live in, whether directly or indirectly. It's also unlikely that many of our parents and caregivers had the necessary tools to regulate themselves, so these traumas were wordlessly passed on to us.

The way we grow up, learn to love and communicate is our foundational imprint. How we show up in the world, both good and bad, is deeply interwoven with who capitalism has allowed us to become. While we can't necessarily change the traumas of capitalism as they are ongoing, rerouting ourselves into healthier more functional ways of responding to it takes slow work and bravery.

In a system that rarely allows us to speak back, or express the full range of our emotions, how are we bringing this into our personal relationships – especially the ones in which we have more power? How can we be safer with one another, and navigate conflict or difficulty in ways that bring us closer to understanding, rather than divide us?

We don't talk about this often because it's too much to reckon with just how much surviving a racist, capitalist society has hardened all of us. It's just as devastating to grapple with what we may have done to others. We atomise our experiences, just as an oppressive culture wants us to and chalk them up to individual histories and failures, without the systemic context in which our personal experiences live.

Being slow enough for the parts of life we've been avoiding can be too tall an order, which is why the lockdowns were a hard time for many. But it was also a time when so many changed and realised something fundamental about their lives. Truths emerged, and despite the rush back to 'normal', the questions that period asked of us and our lives can't be undone. As challenging as many of those questions may be, it's better to engage in the pain of healing than to stay in the wounding of ongoing denial.

This critical self-reflection is one of the crucial ways we can begin to find safety in ourselves, which we can then offer to others. This can look like cultivating enough emotional regulation to be in disagreement with others without shutting down in instances where there may be challenge but no physical or emotional danger. Cultivating personal boundaries and having the courage to stick to them makes it easier to reciprocate, respect and honour the boundaries of others. Safety also looks like acceptance of others, which doesn't hinge on your ability to agree or understand them, but rather on valuing them as human beings.

These are behaviours that are often sorely missing in the world and systems we currently live in. But if we truly want to commit to cultivating different kinds of relationships within our lives and wider community, we're all going to need the presence and safety of others – just as others will need the presence and safety of you.

9.6: Slowing the means of production

CW: workplace exploitation and death

Then there are the working environments where slowing down is impossible: for example, Amazon, where it's been widely reported that people working in their warehouses are often under so much pressure to deliver that they don't have time to even go to the toilet, resorting to peeing in bottles. In such a punitive culture, even breaks and sick leave have reportedly been frowned upon.[1]

In 2022, Rick Jacobs died of a heart attack while working at an Amazon warehouse. Colleagues weren't even notified and the work kept on as usual as he lay cordoned off on the floor.[2] Here lie the very ugliest parts of the system, where there is no pretence of care and consideration, neither in our deaths nor in the basic requirements of our personhood.

In these instances, we can slow down our demand, support strike actions and boycott companies to slow them down from the front end. We are the ecosystem, and we contribute to the pace. This will require a cultural shift in our expectations of instant, around-the-clock service, where instead we buy less and choose more local and ethical alternatives. We control the demand because we are the demand. If we slow down, so does everything else, and the ripple effect continues.

This is already happening, whether corporations like it or not. As of 2024, six in ten adults were spending less on non-essentials due to the cost-of-living crisis.[3] I know I'm buying significantly less these days and being far more mindful when I do have to buy something.

Targeted boycotts also have great capacity to slow demand. In the US, in response to the popular department store retailer Target rolling back its diversity, equity and inclusion (DEI) initiatives in January 2025, organised boycotts – led in particular by the Black community – have led to a steep drop in Target's popularity,[4] with the company reportedly losing an eyewatering $12.4 billion dollars in revenue.[5]

We're also much sicker as a society, contributing less on the supply side via our employment, with life expectancy the lowest it's been in recent years.[6] Not only this, but an estimated of one in five of us is set to have a long-term illness by 2040, with Covid being neglected in recent years almost certainly playing a contributing role.[7] Working under late-stage capitalism is resulting in our sickness. Of course, in a world so deeply ableist, this is covered up with either straight-faced denial or more palatable explanations to cover how debilitated many of us are.

The term 'bed rotting', popularised on TikTok, describes disengaging from daily responsibilities to spend the day in bed or on the sofa, often scrolling on social media, particularly when feeling overly stressed, ill or burnt out.[8] This is hardly novel stuff, but its prominence as a phenomenon chimes with our exhaustion and increasing levels of illness.

There has never been a better a time in our history to learn from disability movements and activists. They've had to develop relationships to time and pace beyond the hyper-productivity ableism demands of us. Crip time is a framework that offers disabled people a more open negotiation of the time and space they need to complete a task. In the words of the academic Alison Kafer, 'Rather than bend disabled bodies and minds to meet the clock, Crip time bends the clock to meet disabled bodies and minds.'[9] A task takes the time it takes.

This way of being might well be the best way to resist the punishing demands capitalism readily expects of us. It's also the reason why the mistreatment of disabled communities continues to be so prevalent – because disability centres human need, not a capitalistic one.

There is much we can learn from this, in the accommodations we ask for – not just at work, but in everyday parts of our lives. By creating this space within us, we can develop ways to understand one another more and expect much less from each other. A task takes as long as it takes.

I hope future generations will come to understand the pace of existing we've normalised as being strange and alien – like the bewilderment many of us experienced when learning about outdated historical norms at school. But in order to achieve this, we must play a part in modelling alternatives to ourselves and each other. Future generations will learn from us, just as we learned from many of our elders how to grind and work twice as hard. That was the survival logic of their times, but ours must evolve – and be very different.

Slowing down is perhaps the most radical thing we can do.

10.

Telling the Truth

*Beware: All too often, we say what we hear others say. We see what we
are permitted to see. Much worse, we see what we're told that we see.
Repetition and pride are the keys to this. To hear or to see even an obvious
lie again and again and again, is to say it, almost by reflex, and then to
defend it because we have said it, and to embrace what we've defended.
Thus, without thought or intent, we make mere echoes of ourselves and we
say what we hear others say*

Octavia E. Butler, *Parable of the Talents*

10.1: Our forgetfulness in a 'post-truth' era (and the bribe)

We must remember again and again because we will forget again and again

Yumi Sakugawa[1]

Years ago, I used to go to a kick-boxing HIIT class at my gym, taught by a woman we'll call Kim. She was a powerhouse; every kick and burpee carried such a level of finesse and strength that we, the class, could only ever clumsily follow behind. She was strong-minded and tough – the kind of person who, if caught in a mugging situation, would make you worry for the other person. For every punch, she helped us conjure our enemies, real or imagined, to target our fists towards. It was the best class.

One session, a figure kept lurking by the window in the door. But we couldn't really see anything – just a shadow that kept coming and going. Eventually convinced that perhaps it was some sketchy guy perving over a class of predominately women, Kim started screaming at the ominous shadow to piss off, flipping the bird for good measure – a lioness protecting her cubs, growling at an unknown predator. Come to find, it was no more than an older woman, who, too shy to come to the class, had been trying to get a sense of it from the window. She complained that Kim had been verbally abusive, even going as far as showing the middle finger.

Kim flat out denied it. *She would never have done such a thing. She couldn't see who was at the door. It had been a misunderstanding. She* was so passionate about denying the whole thing, it became a truth we all quickly fell in line with. Maybe we didn't see exactly what had happened? The music was loud after all, and there was all that adrenaline. Such was the scale of her denial – or our admiration for her – that I think we really believed it, unquestioningly. Not even in the usual changing room chit-chat after class was there any concession.

When ultimately all apologies had finally been made and due process followed, I was left to grapple with how easily our realities are rebuilt in favour of the people we admire, want to protect or who hold any kind of power over us – sometimes without even realising we're doing it.

Oftentimes, the truth is connected to loyalty around who and what we'd ultimately prefer to believe. For this reason, any story is possible.

We live in a time when misinformation, disinformation and alternative facts run rife through our timelines. These days, the truth is only as reliable as our algorithms – yielding to our biases, personal truths and political affiliations. Entire events can be edited and reframed into more palatable narratives. The sweetest person you've ever met could be subject to an online hate campaign, while the good-looking abuser is given an endless redemption arc.

We can believe anything we want to, buried in a perpetual scroll, delving further into our silos. Our capacity to block and scroll past the things that challenge and annoy us, while offering some safety, can also distort our opinions and make us less tolerant of each other in our physical worlds.

While lies and convenient truths have existed since the dawn of time, Trump's initial rise to power in 2016 has ushered in an era where there's no apparent shame to the lie, which has only become cemented by his re-election. It's become entirely possible to launch political campaigns and presidencies built on farce – on fantastical mistruths that, said often enough and with enough power, create a new story, in which the truth becomes an inconsequential hiccup in a freshly pressed message. This has had a ripple effect within society and the world at large. Up is down and down is up. Truth-tellers wanting to cut through the noise must do so amid trolling, censorship and shadow banning – with very little protection in the virtual realm.

Terms are also increasingly being stripped of their history and repackaged in ways that obscure their true meaning. Suddenly, 'femcels' are out to victimise men, racism is a white concern, and 'All Lives Matter'. Ideas that were apparently too difficult to understand when created and applied to the Black struggle are now being readily used in defence of whiteness.

Author Naomi Klein calls this phenomenon 'racial role-playing',[2] where terms, phases and ideas of Black liberation are co-opted, often by controversial figures and movements that morph and stretch these terms until they're rendered meaningless (RIP *woke*). Just like the Spiderman meme of the two identical figures pointing at each other – who is who, when both are wearing the same costume and calling each other the liar with equal conviction?

While this particular shapeshifting of truths is, in many ways, a distinctive feature of our time, rearranging truths and histories is par for the course when it comes to whiteness and its colonial project. Whiteness operates from a place of wilful forgetting. Through a British lens, central to this forgetfulness is the erasure of the most gruesome behaviours, far too unbecoming for the so-called regality Britishness likes to present to the world.

'Operation Legacy' – where thousands of files documenting the full extent of colonial crimes were 'disappeared' by the Foreign Office – was done in the name of forgetting, of wiping the slate clean for fabricated stories fitting a more convenient narrative.[3] At its most absurd, this version of the truth says there is no Black history beyond a very skewed version – one marred by suffering and incompetence. In this world, there are no memories of what was done to us, and racism, in this day and age, is dismissed as a ridiculous accusation – something that happened so far in the past that it's not worth mentioning at all. Your damage and your struggle are yours alone – a question of attitude.

It's why, despite well-documented histories, all evidence and surviving documentation, you can still watch certain TV panel discussions where even basic conversations about racism are challenged and undermined. The Black person on the panel is often pecked away at, while tasked with having to explain the systemic underpinnings of racism in twenty-second intervals – only to be interrupted.

The accepted level of the collective knowledge of racism and how it operates lags far behind its lived reality. This is intentional because it slows our progress. Nothing can move forward if people 'don't know about or understand it' – but they do. Any person living and breathing in this world knows – especially white people who feel threatened by equality. It's an admission of a power they want to hold on to.

There are moments, like the 2020 BLM uprisings, that can break the truth wide open, giving space for a more nuanced public conversation. I think, as a result, the baseline of public conversation around race has shifted. But the quality of our lives, for the most part, remains largely unchanged.

In March 2021, less than a year after the 2020 uprisings, a UK government-commissioned report claimed there was no institutional racism in the UK, and that it should consider itself a model for other white-majority countries[4] – when I first read this statement, I let out such a bitter laugh I surprised even myself. I'm struck by the writer Clint Smith's reflections in understanding the mechanisms behind this kind of denial:

> History is not about primary source documents or empirical evidence . . . It's an heirloom that's passed down across generations. It's something where loyalty to an idea, to a family, to a community, to a sense of self, takes precedence over truth.[5]

The story we're told – the one that's been passed down – is that the UK is a progressive country. That there are boundaries of care around our living – and that the fabric of society will eventually offer all of us a safety net, when and if we need it in our struggles and vulnerabilities. That this is a powerful country that believes in justice and democracy.

This is the script that we are offered and given to remember in place of a far more grizzly truth: that we too often work ourselves to the point of exhaustion and illness only to find that the safety net doesn't exist, only deeper levels of inequality. That we live in a country that has achieved its so-called greatness through consistent acts of global violence and brutality that have contributed to the destruction of our planet, on all levels – environmental, political and social – all of which are largely denied or downplayed.

In a National Public Radio (NPR) podcast interview about James Baldwin, Eddie S. Glaude, author of *The Fire Next Time*, calls the lies we are often called to live by 'the bribe'. He explains:

> The bribe is your silence. The bribe is, you know, just pursue your craft and make money. The bribe is to adjust yourself to injustice. And in the context of the world in which we inhabit, that bribe involves the deformation of our attention, right?[6]

In naming these truths, many of us are faced with defensiveness, minimisation and resentment. Those of us willing to participate in this forgetfulness – who have accepted 'the bribe' – are the first to say things aren't really that bad, or that we need to stop feeling sorry for ourselves. We're supposed to live without context, without history or origin, without feeling. But in choosing to ignore these realities, we leave behind a far more textured and nuanced understanding of ourselves, albeit more painful.

Glaude notes that James Baldwin never accepted the bribe in his life and career. But with that choice came an othering and isolation by the establishment, particularly in his continued support of the Black Panther Party.[7]

Telling the truth in the face of the lie can be dangerous. Many of us know this place all too well – of being further marginalised and punished just for telling it as it is. People lose their jobs, livelihoods and acceptance in the world. But if there are enough of us, the margins we are pushed towards becomes a community and it takes on a whole new positionality. It's on us to do the work of authentically witnessing our lives and by extension, our truths.

I often return to a provocation by Angela Davis: 'Always ask the other question.'[8]

I think that aptly describes where we're at as a larger collective. In our virtual worlds, we've never been more prone to look for other narratives and details beyond the ones we've been given. How many scandals or new stories have been unearthed online – not just by a person talking in the video, but in our increasing capacity to notice what's happening in the background? To follow someone else's POV of the same event? To screenshot a message exchange only shown in passing – and that exchange becoming a story in its own right?

We are developing – at times terrifyingly – an eagle-eyed view of multiple perspectives, and asking other questions to develop fuller truths and answers. While we must be careful to not spread further misinformation and disinformation, there's a potential power we've unlocked here. We are learning to understand the world beyond the single story.

What are you forgetting, and how prepared are you to stand in the truth of what you know once you remember the answers?

10.2: *Tell the truth, shame the devil*

CW: Covid trauma

Growing up in my family, we didn't talk about it. It doesn't pay to be more specific here, because the 'it' in question was all kinds of things; my parents' war, things that felt difficult, things we were hurting from, the things we disagreed with. But the trouble with not speaking about the truth, and the silence it engenders, is that other more powerful things take its place – a more convenient narrative, however far from the truth. It meant that from a distance, my upbringing was smooth sailing, if only you could ignore the torment in the water that existed beneath.

I remember the first time I spoke back to my mother as a young adult – late, all things considered. Her eyes bulged with what I thought then was rage, but on reflecting back on it as an adult, might well have been fear. I broke the seal of silence, I broke the rules, and no one knew what came after. Now, to my personal horror, the world, in its fifth year of the pandemic, holds that same feeling; glossing over and stifling truths for more palatable fairy tales of normality, numbing the reality into obscurity.

Our stance on Covid says much about our collective capacity to tell the truth and look at it squarely. It wasn't that long ago, that *not* wearing a mask was considered anti-social behaviour, with people who refused to do this touted as 'Covidiots'.[1] People who refused the vaccination were also ridiculed, and at times ostracised. That's a far cry from where we are now, where those of us continuing to mask are all but social outcasts, with mask bans even emerging in the US.[2] The fact public opinion can swing like a pendulum to and from such extremes in a number of years should tell us that more than the truths of the pandemic are at play here. It shows how we as a collective can be moulded into enacting the will of public messaging, even when it goes beyond our better interests.

The pandemic is just one example of how the lies of dismissing it as 'over' take us further away from our own understanding and perpetuate a story that doesn't belong to us. In the quest for 'going back to normal', we're not just missing the very real necessities and risks of what it is to be in an ongoing pandemic, but we also leave no space for conversation about what we've lost – just how cruel it was to lose people without having a chance to be there in the hospital wards with them. The worry of people sick with other illnesses, who had their care delayed, or who even put off going to the doctor out of fear of the virus, resulting in an escalation of their ailments.

People who haven't felt the same since their first, second, third infection – and counting. And people who feel OK. The health care workers having to tend first-hand to the lie that Covid is mild – running out of places to put dead bodies, being the last face a dying patient saw before the end. The health care workers with long Covid, who were often working with little to no PPE. The PTSD of it all.

Even the knock-on effects are largely unnamed: students who had their university experience outside of the building, now grappling with the professional world of hybrid working in a cost-of-living crisis. People who were just gaining their confidence in the world, who were pulled back in the lockdowns; the friendships and communities that haven't gained momentum since, or how the National Domestic Abuse Helpline saw a 65 per cent increase in calls from April to July of 2020.[3] The way so much feels changed, while no one wants to talk about it. It goes on.

Positively, too, ideas that were previously seen as impossible happened: the Covid precautions initially put in place, like masking and social distancing, were so effective a strain of flu was eradicated worldwide.[4] The unhoused were given shelter, mutual aid became a more popular term and a US study showed that during the lockdowns (called 'stay in place' in the US) child abuse declined due to the availability of more extended family care, despite fears and previous trends that indicated the opposite would happen.[5]

Online connections were also forged – theatre, music and dance livestreamed into our living rooms; creative ways of connecting that actually included more disabled and chronically ill folks were developed, where previously it was claimed to be unrealistic to do so. All these realities also swept under the rug, like it never happened.

We are missing such a huge amount of analysis, and ultimately deeper truths of who we are, when we undermine Covid as over. The governments' shocking move to cut disability benefits by £5 billion[6] by 2030 is particularly heinous, given both Tory and Labour governments' outright neglect in protecting us from Covid, a disabling virus. In a time when disability is increasing, to cut the support both existing and newly disabled people need is a vicious move.

While collective outrage at these cuts has been well documented, this is an argument that has rarely been made within mainstream conversations. In accepting the lie that Covid is mild and over, we have very little ground to protect ourselves when the consequences of that lie come back to bite us.

The ongoing head-scratching over the increase in school absences since the pandemic[7] and growing concerns over how students are struggling with reading and, more generally, learning comprehension skills[8] in schools is painful to watch. While it's true the lockdowns may have disrupted students' learning patterns, particularly for those who were already disadvantaged, there seems to be little reflection on how the physical impact of Covid is affecting young people in education.

This is a virus that affects brain health – our memory, concentration, our ability to retain information and think critically – even in mild cases,[9] never mind after repeated infections or long Covid. Instead, children are often blamed for their lack of attention, or their parents and teachers are questioned. The truer reality of how we are failing to protect our children is devastating and awful. But to dance around this and deny these truths are much worse.

With millions dead and many more disabled – and counting – there are new, truer stories to tell about ourselves that we are ignoring: culturally, emotionally and socially, even economically. We can't even understand the truth of what's happening to our bodies if we continue to censor conversations around Covid and what it can do to us physically.

And so, when I walk into any kind of space wearing a mask, I see people's eye bulge like my mother's did that day, breaking the rule to participate in the lie of it. I also recognise that for those living by the more widely accepted mistruth that Covid is over, my continuing to wear a mask tells another story, one we've been told is implausible.

Sometimes I see in those eyes something that looks like revisiting a memory – one we've been told we have no right to claim. But rarely, when the truth has been supressed, is the conversation over. It will always find a way to out itself, however long after the fact. If, as many of us have been taught, naming the truth is what will lead to danger, know that living by the lie is worse – for its power to prevent us from deeper, truer learnings.

Naming the truth is the thing that begins to free us.

10.3: Aggression, 'madness' and ridicule

In speaking up or speaking out, you upset the situation. That you have described what was said by another as a problem means you have created a problem. You become the problem you create

Sara Ahmed[1]

CW: ableist language

When the truth can't be refuted, the problem then shifts to its messenger; so we're angry or defensive, said through tears for good measure, especially if the situation is particularly white and gendered. In a professional context, the consequences might look like probation, extra supervision, termination, or an environment so punishing and hostile that we have no choice but to leave.

More than punishing consequence is the denial of our experiences and insights. Through the lens of racism, the truths we know and the things we see all come from a place of imagined inferiority and so can easily be discarded – until a more palatable messenger can say near enough the same thing and be celebrated. Think of any time a cis-gendered white man speaks about domestic violence, racism or misogyny – cue: awe, tears, standing ovation.

There's a confidence that's expected from whoever tells the truth, which is why the white man, perhaps viewed as the epitome of confidence, is more often believed. Uncertainty is, therefore, viewed as an admission of deceit or suspicion, enough for the truth as you know it to be discounted. Never mind that feelings of doubt and shaky confidence are completely understandable responses to having to voice unpopular truths.

Theatre maker and writer Travis Alabanza talks about how uncertainty around gender expression is often weaponised in their book *None of the Above*. For Travis, 'The double standards placed on trans people to be so sure of themselves, confident in our self-knowledge, limit our full humanity and honesty'.[2]

Travis explains that doubts around gender and transness can rarely exist in mainstream culture without being perceived as evidence of some personal failing. While rooted in their personal experiences around gender, this also speaks more broadly to how people with identities that sit outside of perceived norms must operate – knowingly, above all else. Nuance be damned.

Ridicule and mockery are powerful shaming tools against anyone brave enough to go against norms created within the mainstream. People committed to direct actions around the climate crisis have been undermined and deemed as out of touch troublemakers for years. It's also far more expensive and inconvenient to avoid plane travel, eat more mindfully and avoid excessive plastic usage. Anti-racists, still asserting the world we live in is indeed – shock-horror – racist, are told they can't let go of the past, and have chips on their shoulders. Wearing a mask is now almost taboo, with those who wear them dismissed as being overly anxious, and it's near impossible to maintain high-level precautions in a world committed to forgetting Covid ever existed.

Any given products we own are from brands implicated in funding genocides and unlawful working practices, both locally and globally. Being truthful and living truthfully is nearly impossible to do all the time, and so we remain implicated, as surviving under capitalism demands of us. This then becomes weaponised against those trying to call out these truths – that we can't possibly protest the world we're living in, because – what phone are you using? What brand are your trainers? Accusations meant to silence us.

When inconvenience, punishment or mockery isn't enough, along comes the accusation of 'madness'. I use the term 'madness' in inverted commas to problematise its ableist associations and to acknowledge how any perceived deviance from the norm is weaponised and used as a way to undermine and create suspicion.

Nothing in this labelling considers the fact that it is entirely possible to suffer from mental illness *and* be correct – just as people without mental illnesses can be very wrong. Ideas of right and wrong, goodness and badness, have nothing to do with mental health and disability. Not least because living under capitalism is disabling to the Black body in particular – and the rise of mental illnesses in reaction to its pitfalls is an entirely reasonable human response. Being mentally ill doesn't negate having a point – often, quite the opposite. But through an ableist lens and the many cultural depictions of mental illness, it pathologises and undermines the messenger.

History is littered with examples of this. Transphobic hate speech would rather stereotype and paint people as dangerous or mentally ill than understand the complexities and truths of gender identity beyond the binary.

When the physician Ignaz Semmelweis advocated for washing hands in medical settings to prevent the spread of infection in 1850, he was met with scepticism and ridicule from his peers, many of whom deemed him too unstable for the profession.[3]

Semmelweis actually died in a psychiatric institution, though historians still debate the reasons.[4] But being repeatedly gaslighted and told your realities are wrong can lead to a breakdown in mental health, in having to wade through the denial until it overcomes you.

Only a year later in 1851, drapetomania was the term coined for enslaved people who tried to escape the plantation. The bravery, presence of mind and acute sense of their inherent right to freedom were explained away by its author Samuel A. Cartwright, who dubbed these behaviours as mental illness. You see, the lies don't have to be particularly clever, do they?! You just need power, a lie and to say it enough times, and some made-up theory will emerge from the water.

Notions of mental illness and outlier behaviour still follow Black people around, particularly those who deviate from the script of normality, a standard in and of itself devised by whiteness. Our reasonable objections to having to navigate racist behaviours are labelled as aggressive and angry. The hyper focus on us and our behaviours reveals the commitment to keep seeing Black people as non-human. But, as per Angela Davis's steer, if we ask the other question, it would be why whiteness would go to such depths to point the finger and create myths around us while never interrogating itself – such as power rarely has the inclination to do.

In these increasingly fascist times, what is 'normal' and deviant are policed and enforced loudly. Fear is the objective – so, wearing a mask, talking about race, objecting to genocides, protesting, being vocal and direct about the climate crisis, all come with increasing consequence, especially depending on who is doing the talking.

To live defiantly by these ideals will at times feel thankless – because it's supposed to. Little to no reward is offered for non-compliance of any kind. This is where those with relative privilege must step to the front, particularly those with financial security or proximity to whiteness. Privilege is a shield, and those who can stand behind it are more likely to be seen as eccentric or just alternative when speaking out than be punished or brutalised.

Regardless of where we are in the scheme of things, we will need to live by truths that stand defiantly against many of the norms we have been taught to live with. We must speak up when it is difficult – or support others to do so. Both scarily and fortunately, our norms are rapidly disintegrating, and change is happening in quick bursts. In all of this, the truths we are brave enough to acknowledge and speak aloud must stand the test of these times.

10.4: We were never meant to survive

So it is better to speak
remembering
we were never meant to survive

Audre Lorde, 'A Litany for Survival'

I write all this to say, speak, even when you are unsure and it's difficult.

I have a printout of Audre Lorde's poem 'A Litany for Survival' on my wall. I have it there as a reminder, particularly for times of my life when speaking the truth has felt difficult.

In it, she speaks to 'those of us standing upon the constant edges of decision / crucial and alone' and those of us 'who were imprinted with fear'.

What I love about coming back to this work – as I continue to do – is that for Audre, fear is not the final destination. It's not the end of the road, but rather a way we move through the world, fear and uncertainty in hand, with the wisdom they offer. It sits in parallel to her other words:

> My silences had not protected me. Your silence will not protect you. But for every real word spoken, for every attempt I had ever made to speak those truths for which I am still seeking, I had made contact with other women while we examined

the words to fit a world in which we all believed, bridging our differences.[1]

We've all been in rooms and places that don't want us there; we've all felt what it is to be the 'only one'. I don't want to romanticise how hard that is, but there is an alchemy and power in our words, forever seeking those who need them, and building new worlds. There is beauty in the vulnerability of speaking words – shaky and uncertain – that, without you, would have otherwise been unsaid, never meant to survive.

The ending of an empire marks a shift of the world order as we know it, and by some shitty miracle we find ourselves here, living through it. I say both 'shitty' and 'miracle' because, at least from my perspective, it's both. Shitty because any change on this scale, any concession of power and the order of things precedes chaos. Those of us living and surviving in the imperial core know its callousness and its cruelty intimately. We also know how the most vulnerable of us will be most affected in the wake of its struggle to maintain its power.

But even within this, I hold space for the miracle of witnessing it, the living proof that even the most immovable of things can change. We are changing into something else and with it comes the possibility, even obligation, to also become transformed.

Rarely does change come with a neat beginning, middle and end. It may well be that the end of this empire drags itself through most of our lifetimes. But with its hold loosening, comes the potential for more of us to critique and divest of the institutions that prop up its power.

Here is a greater potential for us to do the very opposite of what these systems believe in: resource hoarding, living in isolation and enacting a global racial and ableist order. We have to change the default many of our current norms were built upon. The truth is the realest foundation we have to build upon, one from which something much much better than this can emerge.

PART FIVE

How We Thrive

All I know is I'm getting further and further away from the plantation that owned my body, and claimed my labour, and owned my mind, and insisted on me thinking in particular ways

Bayo Akomolafe[1]

I want you to remember that most things are an invention

Lola Olufemi[2]

This is a chapter of uncertainty, of untidy answers, non-endings and non-linear thinking.

Part of the work of the writer is reimagining the futures we're heading towards, using words to help heal past wrongs and set us on a clearer path towards them. But creativity can't be rationed to only a few of us, it belongs to everyone. A large part of this chapter is gifting your own creativity back to you as something you have always had, too often undervalued and unloved in our current reality.

The different sections of this essay take inspiration from different thinkers of radical imagination. Because simply put: I am unsure. Writing about our struggles and what's difficult comes much easier to me, but it's not a place I want to stay in. So, I take guidance from others to lead us through ways we can all envision the future we are walking towards.

This is an exercise in collective imagination.

Radical Imagination

11.1: Leaving the white man's imagination – Lessons from Ayanda

We are living within the imagination of white, colonial men who decided what it means to be beautiful, what it means to be worthy, what it means to be productive. Why is productivity even a value that we have? Who does that benefit?

Ayandastood[1]

We're living within the imagination of whiteness. Pack it up, because there is nothing here for us, nor will there ever be. Everything from our labour, our beauty and our ideas has been exploited, commodified and stripped of their worth and meaning by design.

The climate crisis speaks to the consequences – whether wilful or unintended – of this manufactured way of thinking, which whiteness has had the power to execute and make real. The rest of us surviving within this world are too keenly aware not just of its cruelty and make-believe racial order but also of its ignorance and its shortcomings. That these ways of being – of living unsustainably, extracting the world of all its resources and being out of sync with nature – ultimately can't go on forever.

Understanding the fiction our whole lives are built upon means that the way to undo this also starts with the mind. It means we can make up our own ideas, live by our own values and debunk the idea of a bigger power knowing what's best for us – particularly when that power means to subdue and ultimately harm us.

If this makes you uncomfortable or angry, then its an indication of how much you still have a stake, real or imagined, in this system continuing as it is. But for people of conscience, we know that any system that has the power to violently subjugate and demean *anyone* is a dangerous system not worth investing in at all.

The trauma of living within this system inherently dulls our capacity to dream because to survive, we've often had to cultivate an entirely different set of tools grounded in this reality. But we need to tend to our innate capacity to reimagine and to look after it. We need to work through the falsehoods we've been told about not being creative or good enough, and make space to create our own ways of being, based on our deepest understanding of worlds we know have the right to emerge.

Our imagination is also a territory that has been stolen from us, and there is often very little space for our own thoughts and aspirations. From childhood, we've been told by society what our lives are for and the norms and customs to live by: education, job, marriage, children, work till retirement. These have been life's preordained milestones, and they've loomed so large that there's often been very little space to dream of much else, beyond achieving and upholding them. But so many of us are questioning the nature and obligation of all these things – fewer of us are having children or getting married, and many more of us are questioning our relationships with work. These are attitudes that were inconceivable to the mainstream only a few decades ago.

In these times of rapid societal change, it's important we know our minds and our values. Much like land that has been co-opted by someone else, we must tend to our minds with a farmer's love – excavating the weeds that don't belong there, tending to the soil and taking time to replant the space with things that replenish and heal.

If we don't take the time to do this, we can very easily get lost in other people's vision and ideas for the future – ones that replace exploitation with more exploitation. What does life look like for us beyond the cruelties of late-stage capitalism? Who would you want to be? Speak to these ideas, ask them to come towards you and make sure you're ready to hear their invitation.

11.2: Be unrealistic – Lessons from Walidah

We live in capitalism. Its power seems inescapable. But then, so did the
divine right of kings. Any human power can be resisted and changed by
human beings

Ursula K. Le Guin[1]

Big ideas can often be dismissed as frivolous and unrealistic,
too disconnected from our current reality to deserve serious
consideration. This is something I've thought a lot about in
writing this book, which contains many ideas for which I have
no clear sense of how they will come into being.

How realistic is it to talk about open borders or abolishing
the police in a world devoted to holding on to these systems,
actively reinforcing and expanding them? How much hope can
we have, given the oppressive forces on the planet that consistently choose death and devastation over our lives?

It's rational to understand these as unrealistic ideas within
this current climate, and I think we have to hold space for the
grief caused by these harmful choices that are being made
around us, over and over. It's, understandably, all too easy
to get caught up in the shape of our realities and lose sight
of the bigger picture we are mere specks within. We may
never see the changes we long for in our lifetime, but we
are carrying on a legacy of daring to believe things can be
better, so the generations coming after us have something to
build from.

James Baldwin once said, 'You think your pain and your heart-break are unprecedented in the history of the world, but then you read.'[2] Our elders, in books, prayer, songs and memory, have shared their pain and their lessons. In ways both conscious and subconscious, they move with us. What kind of lessons do we, too, want to leave behind for those on their way?

We need to always be pushing the needle of what we and future generations deserve, leaving those desires, dreams, ideas and lessons with them, because maybe for future generations many of these ideas will be possible. Just as we are living some of our ancestors' dreams that were once impossible. We must leave a blueprint.

Consider this as being a part of an inter-generational, inter-dimensional conversation, happening back and forth and around, beyond time and linear thinking.

This is not to gaslight ourselves or ignore the very real difficulties that we currently face, but within it all to cultivate a clear space in our minds – a place protected from 'what ifs' and 'buts' – to allow us the space to dream of something else. This is the most powerful inheritance we can gift to others and ourselves.

For some writers, sci-fi is a genre through which they can explore so-called unrealistic ideas – creating new fictional worlds where concepts of racism, or any other kind of oppression, have no meaning. This has also given space for Afrofuturism to emerge – an art form that merges Black mythologies and histories with sci-fi and imagined futures to explore Black agency and liberation.[3]

The writer and activist Walidah Imarisha advocates for sci-fi – or in her terms, visionary fiction – as a means to re-envision justice and talk about a world without prisons and police violence, capitalism or any of the oppressive systems that stop us from living full and meaningful lives.[4]

In her words, visionary fiction reminds us to be 'utterly unrealistic in our organizing, because it is only through imagining the so-called impossible that we can begin to concretely build it'.[5] The premise of writer Cebo Cambell's novel, *A Sky Full of Elephants*, for example, is a world where all white people have disappeared in America, offering a new reality where the protagonists must figure the world out for themselves, without racial oppressions and binaries. For Walidah, also co-editor of the visionary fiction anthology *Octavia's Brood: Science Fiction Stories from Social Justice Movements*, 'When we free our imaginations, we question everything. We recognize none of this is fixed, everything is stardust, and we have the strength to cast it however we will.'[6]

Ending four hundred years of transatlantic slavery was once an unrealistic idea, as was the fall of empires and political systems – good and bad – we now understand to be in the past. They once loomed as large and powerful as the one we're surviving in right now. In the thick of those times, when all seemed lost, all that was left was the mind.

Understanding our lives beyond these worlds demands we invent new ones. It's deceptively simple, and thinking this way can feel inherently unsafe. But this is magic work, something that children are naturally very good at; as are, bizarrely, the far right. This is something that Naomi Klein touches on in her book *Doppelganger* in which she details the ways the far right's disinformation and conspiracy theories have brought their ideas into public consciousness – whether we like it or not. The levels of utter nonsense paraded as fact right now are staggering. And as damaging as the rise of disinformation continues to be, there is also something we can learn from in this – ideas are powerful, and to speak them aloud again and again is to bring them closer into the possibility of existence.

It's worth mentioning that I struggle with this massively. As a researcher, I look to data, evidence and set parameters from which to tell a story. It's the poet in me that drags me towards the possibilities, but it's not always a natural stopping point for me. More often, it feels like being asked to do a trust fall – with no clue whose hands you're falling into.

It can be hard to see the light at the end of the tunnel when so much seems to be heavily oppressive. But something I've recently learned is about the role of radical optimism within the Palestinian struggle. It's considered a religious and spiritual obligation (known as a wajib) to commit to the daily practice of optimism, of faith in Palestinian liberation.[7]

It's also why it's important to practice optimism in community, first, so we develop a collective idea of liberation, but also precisely because it can be so hard to believe when faced with our current realities. It's hard work to look at the immoveable nature of things and believe they can change for the better. But dreaming with others and co-creating alternatives of the mind hold a certain kind of alchemy that can often be just as unshakeable.

Life often has a trickster way about it, and the changes we've been longing for and working towards can happen often when we least expect it. What we can imagine builds the path we will walk in the future, and how we think will usher in the shape of newer worlds to come.

11.3: *Someone who loves me – Ordinary utopias – Lessons from Reagan*

What the world will become already exists in fragments and pieces, experiments and possibilities

Ruth Wilson Gilmore[1]

Utopia is in the ordinary.

The extent of what was taken from us means we can often think of liberation as complicated, and in the legacy of visionary fiction, it sometimes means we have to create other worlds in order to conceive of realities beyond the here and now.

While the means of achieving liberation may well be difficult, the ultimate outcome is very simple: living a fulfilled life of respect, love and dignity, in whatever shape we each choose, causing no harm to others.

When I wake up there is someone who loves me.

This is the first line from writer and artist Reagan Jackson in her contribution to Natasha Marlin's anthology, *Black Imagination*.

When I leave my home, the people living on my street know my name . . . we break bread together. We laugh and dance and build together,[2] Jackson continues.

Simple utopias.

For those of us coming from a family history of colonial violence, war and migration, we are here because of someone in our lineage and their fierce tenacity to survive. Our families have had to fight to have a place in this world and imagine our lives within it. As many of us are all too familiar with, living in ways that are self-defined often creates a tension that bristles and disrupts white-supremacist thinking – a system that can only imagine our servitude and inferiority.

But within this tension we've honoured and built our own rituals, spiritual and religious practices, communion and collective care. It's precisely in our capacity to create and maintain these things that we've been sustained alongside the racial terror and oppression of life within the imperial core. In the unravelling of life as we know it, worsened by the climate crisis, it's these very ordinary things that we must hold on to – simple life-giving joys.

What rituals and reminders can we create, as did those who came before us, to hold us afloat? This is not to completely romanticise our elders. Many of us know first-hand that with their survival came their silence and unhealthy emotional coping patterns. But how do we learn the right lessons from them?

I'm trying to understand how we hold dualities.

When I think of ordinary utopias, I think of washing the dirt off a bundle of veg gifted from a friend's garden, the aftermath of a difficult conversation with a friend where there's been a will to listen and learn from one another. I think of liking the person I see in the mirror, making myself laugh, no competition, random bouts of kindness, learning how to be a safe person to myself and others, gatherings with other Black folk, lots of rest and play, times when I just can't do it, and being okay with it, and a deep acceptance of myself and those around me.

While some of these things aren't always available to me under capitalism, many are, and mostly can be the result of my choices and the personal parameters that I set for my life. They can be found in the relationships and communities I want to cultivate, and the personal growth and emotional work I'm willing to commit to.

I encourage you to develop your own list and see how many of the things you come up with are within your power and courage to achieve now, rather than waiting for some greater change to come. The change is us and how we start to live in the here and now. What rituals and ordinary utopias can be made in the spaces we inhabit? How can we honour the rituals of the past we grew up with in new ways that sustain us for our current times?

11.4: God is change – Lessons from Octavia

The only lasting truth is change

Octavia E. Butler, *Parable of the Sower*

I don't know about you, but some of the hardest times of my life were made significantly worse by my reluctance to let go, by my sheer determination to hold onto things as I thought they should be, not as they actually were. For most of my life, I've wanted to know exactly what is happening and when. But the ongoing uncertainties of more recent years have been a rude awakening, and have pushed me into the weary acceptance that certainty is an illusion and one that, now more than ever, we don't have the luxury of believing in.

It can be hard to look into the vast unknown that change offers, but as the climate crisis is teaching us, nothing stays the same. That is a promise. As much as change requires preparation and pre-empting, it also requires our fluidity and our ability to pivot and adapt. The pain of the climate crisis is precisely because of the slowness of the powerful and political elites to adapt to the changes required of us, only doing so when it is financially and politically viable, and at the expense of so many of us.

Writer Octavia E. Butler was a visionary in the works she created, and the lessons of her work reverberate loudly through these times. In her book *Parable of the Sower*, we are taken to an apocalyptic world marred by climate change and deep inequality. Within this, we are shown how people who are reluctant to change don't survive, while Lauren, the main protagonist, adopts the mindset that God is Change. She does this at odds with her family and community, as they attempt to cling onto some semblance of normality.

I recognise this – the cult of normality that plagues us all. Life has hard-wired us to 'keep calm and carry on', despite ourselves and the state of the world. The book, published in 1993, and eerily set in 2024, offers us many lessons to hold onto. Often, we cling to the familiarity of old habits and belief systems even when they're hurting us. But as we live through the very timeframe Octavia's book was set within, we'd do well to listen to the warnings her work is offering us.

We are living through immense amounts of change, flitting between our fears of what might happen while living them. The dangers around the climate crisis we fear are already here and being experienced in several parts of the world simultaneously. The rise of the far right and fascism are here, as is ongoing racial hatred and corrupt political action. No one is going to ring the bell and say it's arrived. It's only the hindsight of history that will attempt to label what's happening in clearer, distinct terms, but right now, it's happening in spikes and shocks – sometimes suddenly and with great intensity, and at other times simmering quietly. However, so is our resistance and our bravery.

Like Lauren, we have a decision to make about who we're going to be during this time, and what ways of being we will adopt, even when it is seemingly counter to the dominant culture around us. One thing is for sure, inaction is still a decision;

so whether we are inactive or proactive during this time, it's unlikely we will remain unchanged. Worlds have already ended.

For the writer and thinker, Alok Vaid-Menon:

> The universe and the earth templates change as the natural orientation of the world . . . every single example that is templated for us, is not just that change happens, but actually that change is absolutely necessary . . . the unknown is how we've birthed everything. The unknown is actually a fertile ground to build the world that we dream of.[1]

Just as the planet is changeable by its very nature, so too are we as a part of its ecosystem. We change and adapt as a means of our survival. Our systems and cultural norms are disintegrating; people in power make less and less sense, and our governments are woefully ill-prepared for and unresponsive to the gravity of the challenges we face. The less we build our lives around ideas that are sinking, the better. That way when they fall, we won't go with them. This is where our marginalisation might save us because there is less detaching we have to do in a system that has never really included us.

The lives and communities we deserve are on the other side of the changes we are resisting. Now is the time to be honest with yourself: what are you willing to let go, and what unknowns are you willing to work towards? How might you discover new parts of yourself in the transformation?

Concluding Note

No one is coming to save us.

This is something that, as Black people, we should already know by now. There has been no difficulty, deep trauma or level of pain we've endured where we haven't had to tend to ourselves and be our own saviours. I've struggled and I've grieved this for years, but who else should do it? Who else but you and I should hold the task of shaping our own existence?

The future we're heading towards will require more of us to act in community and solidarity with one another than ever before. I think the most important thing we can also do is rebuild trust among one another, which is needed to forge deep community bonds. This also means a reckoning with how we have been harmful and committing to doing differently, beyond words and through changed behaviours and actions.

Trust within ourselves will also be important. In a time when much divides us and disinformation is widespread, it's important that we're able to develop an inner compass, helping us to navigate our way through.

Here are some guiding questions:

- Who do I believe and why?

- Who and what wants me alive and well? (And who wants my servitude?)
- What has outstayed its welcome in my life?
- Who am I in this moment? What systems are my behaviours supporting?
- Who are the 'masters' living in my head? What steps am I taking to show them the door?
- Whose permission am I waiting for?

Consider these questions as starting points, not definitive conclusions. A roadmap to another beginning.

Within this book, I have tried to connect some of the factors – environmental and otherwise – that influence and shape our lives. But the reality is, much more is approaching us. Many of us may be able to anticipate what is to come, recognising the warning signs of history repeating itself. It is likely there will be continued deep societal division and extremism, violence stemming from all-time-high levels of fascism and eugenics and the scarcity of resources and global conflict. But there will also be many more unknowns, and much more to discover in how we choose to respond.

This will be our next chapter, and it's on us, both individually and collectively, to respond to what it asks of us, seeking to resist exploitative systems that strive to violently harm us. It's on us to look out for one another, to be able to sift through the noise and reclaim the values and ideals that matter to us most. To rebuild community and ways of living from our truest image, finding our way back to the land, ourselves and each other.

Acknowledgements

Many people in ways big and small have helped bring this book into existence. My particular thanks to:

Julia Oertli, for your encouragement and loving challenge across the various phases of the book, and for being there for the highs and lows.

To Désirée Reynolds, I, like many others, am all the better for your generosity and expansive knowledge as a writer. Thank you for telling me to write this book as if I were talking to you, and for not holding it against me (too much) for only taking that advice after I'd exhausted myself . . . a year later. May the flowers you deserve blossom in their multitudes.

To Hette Phillips, for our setting the world (and this book) to rights, for your solidarity and ideas. For gifting me the line: 'Utopia is in the ordinary.'

To Louise Foreman, Miri Schoop, Nico Yvan, Maya Good-fellow and Grace De Morgan, for hearing some of my very first ideas and drafts, when they had no certain place to land, and for your interest and support.

To Sarah Hymas for holding such a sacred space on 'Imag-inarium Earth', which allowed me to be in the messiness of drafting this book, and to understand it from all angles.

Additional thanks to my fellow participating writers for your encouragement and feedback.

To Phoebe Yang for your care and attention – you may well have saved my life on more than one occasion.

To my agent Crystal Mahey-Morgan and the OWN IT! team, for championing my ideas and this book, way before I knew what it could become.

To my original editor, Charlotte Humphrey, for not only seeing the potential in this book, but also in me as a writer, and to the editorial team past and present, Poppy Hampson, Kaiya Shang and Sam Stocker. To Vimbai Shire for your care and enthusiasm for this book, and to Binita Roy.

A special thanks to everyone I interviewed who gave their time, energy and experiences to help inform this book, in chronological order:

Jennifer Uchendu, Babitha George, Catherine Baxendale, Sylvia Kokunda, Rosamund Adoo-Kissi-Debrah, Angela Fonso, Josh Artus, Simmone Ahiaku, Panagioti Tsolkas, Adam Elliott-Cooper, Sanaz Raji, Stephanie Davis, Hazel Sealeaf, Noora Firaq, Josina Calliste, Evie Muir, Valerie Goode and Cherrelle Douglas.

This is the 'village' this book was raised in, and I am deeply grateful for it.

Notes

Introduction

1 Richard Seymour, *The Disenchanted Earth: Reflections on Ecosocialism and Barbarism* (London: The Indigo Press, Mood Indigo, 2022), 13.
2 Daniel Macmillen Voskoboynik, 'To Fix the Climate Crisis, We Must Face Up to Our Imperial Past', *openDemocracy*, 8 October 2018, https://www.opendemocracy.net/en/opendemocracyuk/to-fix-climate-crisis-we-must-acknowledge-our-imperial-past/.
3 United Nations, 'What Is Climate Change?', https://www.un.org/en/climatechange/what-is-climate-change.
4 European Parliament, 'Climate Change: The Greenhouse Gases Causing Global Warming', 16 March 2023, https://www.europarl.europa.eu/news/en/headlines/society/20230316STO77629/climate-change-the-greenhouse-gases-causing-global-warming.
5 United Nations, 'The Climate Crisis – A Race We Can Win', https://www.un.org/en/un75/climate-crisis-race-we-can-win.
6 Sami Schalk, *Black Disability Politics* (Durham: Duke University Press, 2022), 3.
7 Toni Morrison, quotation from 'Black Studies Center Public Dialogue, Part 2', *Special Collections: Oregon Public Speakers*, Portland State University, 30 May 1975, audio recording available at https://soundcloud.com/portland-state-library/portland-state-black-studies-1.
8 Britt Hawthorne, 'Who Are People of the Global Majority and Why It Matters', *Britt Hawthorne* (blog), 27 March 2023, https://britthawthorne.com/blog/people-global-majority.

Notes

Part One: Air

1. Feeling Our Way Through the Climate Crisis

1.2 Whose crisis, whose timeline?

1 Arinze Chijioke, 'Niger Delta Oil Spills Bring Poverty, Low Crop Yields to Farmers', *Al Jazeera*, 9 September 2022, www.aljazeera.com/features/2022/9/9/niger-delta-oil-spills-bring-poverty-low-crop-yields-to-farmers.

2 International Fund for Agricultural Development (IFAD), 'East Africa Is Experiencing Its Worst Drought in Decades. It's Time to Invest in Climate Adaptation', 2022, https://www.ifad.org/en/web/latest/-/east-africa-is-experiencing-its-worst-drought-in-decades-it-s-time-to-invest-in-climate-adaptation.

3 ReliefWeb, 'Unprecedented Drought Brings Threat of Starvation to Millions in Ethiopia, Kenya, and Somalia', 9 June 2022, https://reliefweb.int/report/ethiopia/unprecedented-drought-brings-threat-starvation-millions-ethiopia-kenya-and-somalia-issue-date-9-june-2022.

4 Friends of the Earth, 'Eco-Anxiety: How to Turn Fear into Hope', https://friendsoftheearth.uk/climate/ecoanxiety-how-turn-fear-hope.

5 Christopher B. Obida et al., 'Quantifying the Exposure of Humans and the Environment to Oil Pollution in the Niger Delta Using Advanced Geostatistical Techniques', *Environment International* 111 (2018), https://www.sciencedirect.com/science/article/abs/pii/S0160412017313193.

6 Friends of the Earth, 'Eco-Anxiety'.

7 Nicole Dennis-Benn, 'Swimmer', in *The Good Immigrant USA: 26 Writers Reflect on America*, ed. Nikesh Shukla and Chimene Suleyman (New York: Little, Brown and Company, 2019), 27.

8 Sarah Jaquette Ray, 'Climate Anxiety Is an Overwhelmingly White Phenomenon', *Scientific American*, 21 March 2021, https:// www.scientificamerican.com/article/the-unbearable-whiteness-of-climate-anxiety/.

9 The Wildlife Trusts, 'High Speed Rail (HS2)', *The Wildlife Trusts*, https://www.wildlifetrusts.org/hs2.

10 You can read more about this in: Elizabeth Schmidt, 'Hands Off Africa', *Africa Is a Country*, 10 February 2023, https://africasa-country.com/2023/02/hands-off-africa.

11 Patrick de Marie Katoto, 'Environmental Threats and Respiratory Health', *CEGEMI*, https://cegemi.com/index.php/environmental-threats-and-respiratory-health-in-kivu/.

12 Oluwole Ojewale, 'What Coltan Mining in the DRC Costs People and the Environment', *The Conversation*, 29 May 2022, https://theconversation.com/what-coltan-mining-in-the-drc-costs-people-and-the-environment-183159.

13 Peer Schouten et al., 'Our Climate Future Depends on Conflict Dynamics in Congo', *Danish Institute for International Studies*, 29 November 2022, https://www.diis.dk/en/research/our-climate-future-depends-on-conflict-dynamics-in-congo.

14 SustyVibes, 'What Can We Learn About Your Experience of Eco-Anxiety?', *The Eco-anxiety in Africa Project (TEAP)*, https://www.teap.sustyvibes.org/.

15 Charles A. Ray, 'The Impact of Climate Change on Africa's Economies', *Foreign Policy Research Institute*, 29 October 2021, https://www.fpri.org/article/2021/10/the-impact-of-climate-change-on-africas-economies/.

16 Shelot Masithi, 'Ubuntu Environmentalist', *Ubiquity University Community*, https://community.ubiquityuniversity.org/profile/shelot-masithi.

17 ONCA, 'Interviews: The Eco-Anxiety Project Africa', May 2022, https://onca.org.uk/interviews-the-eco-anxiety-project-africa/.

18 Ibid.

1.3 (Black) Solastalgia and slow violence

1 Minority Rights Group International, 'Batwa in Uganda', July 2018, https://minorityrights.org/minorities/batwa/.

2 Fred de Sam Lazaro and Sarah Clune Hartman, 'Uganda's Batwa Tribe, Considered Conservation Refugees, See Little Government Support', *PBS NewsHour*, 21 October 2021, https://www.pbs.org/newshour/show/ugandas-batwa-tribe-considered-conservation-refugees-see-little-government-support.

3 Minority Rights Group International, 'Batwa in Uganda'.

4 de Sam Lazaro and Hartman, 'Uganda's Batwa Tribe'.

5 Leon Gooberman, 'The State and Post-Industrial Urban Regeneration: The Reinvention of South Cardiff', *Urban History*, 2018, 45 (3): 504–23, https://doi.org/10.1017/S0963926817000384.

6 Chris Sullivan, 'Lost Cities: How Cardiff's Thriving Multicultural Hub Was Crushed', *Byline Times*, 5 August 2020, https://bylinetimes.com/2020/08/05/lost-cities-how-cardiffs-thriving-multicultural-hub-was-crushed/.

7 Ibid.

8 Thomas Deacon, 'The Inconvenient Truth About Cardiff Bay: The History of Cardiff's Waterfront and How Wales' First BAME Community Was Driven Out to Build It', *Wales Online*, 19 July 2020, https://www.walesonline.co.uk/news/wales-news/inconvenient-truth-cardiff-bay-history-18402792.

9 Sullivan, 'Lost Cities'.

10 Deacon, 'Cardiff Bay'.

11 Ibid.

12 Gooberman, 'The State and Post-Industrial Urban Regeneration', 504–23.

2. Air Pollution

2.1 We can't breathe

1 MedlinePlus, 'Air Pollution', *MedlinePlus*, https://medlineplus.gov/airpollution.html.

2 World Health Organization, 'Air Pollution', *World Health Organization*, https://www.who.int/health-topics/air-pollution#tab=tab_1.

3 Helena Horton, 'Net Zero Climate Strategy: UK Government Sued Over Lack of Action on Air Pollution', *The Guardian*, 12 January 2022, https://www.theguardian.com/environment/2022/jan/12/net-zero-climate-strategy-uk-government-sued.

4 Zawn Villines, 'Air Pollution and Pregnancy Outcomes: What Are the Effects?', *Medical News Today*, 30 October 2020, https://www.medicalnewstoday.com/articles/air-pollution-and-pregnancy-outcomes#air-pollution-and-pregnancy.

5 World Health Organization, 'New WHO Global Air Quality Guidelines Aim to Save Millions of Lives from Air Pollution', 22 September 2021, https://www.who.int/news/item/22-09-2021-new-who-global-air-quality-guidelines-aim-to-save-millions-of-lives-from-air-pollution.

2.2 Somewhere to live – A brief history of air pollution

1 Laurel A. Royer, 'Environmental and Human Health Justice: A Call To Greater Action', *Integrated Environmental Assessment and Management*, 020, https://setac.onlinelibrary.wiley.com/doi/10.1002/ieam.4582.https://setac.onlinelibrary.wiley.com/doi/10.1002/ieam.4582.

Notes

2   Kelly Woolford, '10 Facts You Didn't Know About Streatham', *My Streatham*, 5 February 2019, https://www.mystreatham.com/10-facts-you-didnt-know-about-streatham/.

3   Josh Mcloughlin, '15 Reasons to Go to Streatham High Road, SW16', *Time Out*, 24 March 2017, https://www.timeout.com/london/blog/15-reasons-to-go-to-streatham-high-road-sw16-032417.

4   History.com Editors, 'Industrial Revolution', *History.com*, https://www.history.com/topics/industrial-revolution/industrial-revolution.

5   Kirsten W. W. L. Haeger, 'Air Pollution in Victorian-Era Britain: Its Effects on Health Now Revealed', *The Conversation*, 13 February 2018, https://theconversation.com/air-pollution-in-victorian-era-britain-its-eff-ects-on-health-now-revealed-87208.

6   History.com Editors, 'Industrial Revolution: Negative Effects', *History.com*, https://www.history.com/news/industrial-revolution-negative-effects.

7   Haeger, 'Air Pollution in Victorian-Era Britain'.

8   Ryan Juskus, 'Sacrifice Zones: A Genealogy and Analysis of an Environmental Justice Concept', *Environmental Humanities* 15, no. 1 (1 March 2023): 3–24, https://read.dukeupress.edu/environmental-humanities/article/15/1/3/343379/Sacrifice-ZonesA-Genealogy-and-Analysis-of-an.

9   Greenpeace, 'Waste Incinerators and Deprivation: A New Map Shows the Link', *Unearthed*, 31 July 2020, https://unearthed.greenpeace.org/2020/07/31/waste-incinerators-deprivation-map-recycling/.

10 Kendra Pierre-Louis, 'Past Racist Redlining Practices Increased Climate Burden on Minority Neighborhoods', *Scientific American*, 27 July 2020, https://www.scientificamerican.com/article/past-racist-redlining-practices-increased-climate-burden-on-minority-neighborhoods/.

2.3 Something in the air – Rosamund and Ella Adoo-Kissi-Debrah

1 Claire Marshall, 'Rosamund Adoo-Kissi-Debrah: "Did Air Pollution Kill My Daughter?"' *BBC News*, 29 November 2020, https://www.bbc.co.uk/news/stories-55106501.

2 Richard Morris, 'The Victorian "Change of Air" as Medical and Social Construction', *Journal of Tourism History* 10 (2018): 1–17, https://doi.org/10.1080/1755182X.2018.1425485.

3 Spotlight, 'Time for Change | A Spotlight and Poplar HARCA Film', *YouTube* video, https://www.youtube.com/watch?v=5VuEQ6A3MhY&t=404s.

4 Poplar HARCA, 'New Climate Change Film Tells Tower Hamlets to Act Now', 10 March 2021, https://www.poplarharca.co.uk/about-us/news/article/new-climate-change-film-tells-tower-hamlets-to-act-now/.

5 Sandra Laville, 'Air Pollution in Birmingham "Shortens Lives of Children by Half a Year"', *The Guardian*, 8 July 2019, https://www.theguardian.com/environment/2019/jul/08/air-pollution-in-birmingham-shortens-lives-of-children-by-half-a-year.

6 UK100, 'Evening Roundtable: Air Quality and Communities of Colour', *YouTube* video, 20 January 2021, https://www.youtube.com/watch?v=QnSviEMm6kQ.

7 World Health Organization, 'More Than 90% of the World's Children Breathe Toxic Air Every Day', 29 October 2018, https://www.who.int/news/item/29-10-2018-more-than-90-of-the-worlds-children-breathe-toxic-air-every-day.

2.4 Are you meditating enough?

1 Harvard T.H. Chan School of Public Health, Center for Climate, Health, and the Global Environment, 'Coronavirus and

Pollution', https://www.hsph.harvard.edu/c-change/subtopics/coronavirus-and-pollution/.

2 Ibid.

3 BBC News, 'Covid: Bus Drivers "Three Times More Likely to Die" Than Other Workers', 19 March 2021, https://www.bbc.co.uk/news/uk-england-london-56455845.

2.5 *The trouble with Southall*

1 'Southall Broadway Among London Diesel Pollution Hotspots', *Ealing Times*, 19 August 2020, https://www.ealingtimes.co.uk/news/18661173.southall-broadway-among-london-diesel-pollution-hotspots/.

2 Vivek Chaudhary, 'How London's Southall Became "Little Punjab"', *The Guardian*, 4 April 2018, https://www.theguardian.com/cities/2018/apr/04/how-london-southall-became-little-punjab-?CMP=share_btn_tw.

3 OCSI (Oxford Consultants for Social Inclusion), 'What Can Census 2021 Tell Us About Ethnic Diversity in England?', November 2022, https://ocsi.uk/2022/11/30/what-can-census-2021-tell-us-about-ethnic-diversity-in-england/.

4 'Southall Gas Works', *Wikipedia*, https://en.wikipedia.org/wiki/Southall_Gas_Works.

5 Southall and Hayes Clean Air, 'About', https://southallandhayescleanair.org.uk/about/.

6 Jo Griffin, video by Maeve Shearlaw and Kyri Evangelou, 'Londoners Claim Toxic Air From Gasworks Damaging Their Health', *The Guardian*, 27 August 2020, https://www.theguardian.com/environment/2020/aug/27/londoners-claim-toxic-air-from-gasworks-damaging-their-health.

7 Southall and Hayes Clean Air, 'About'.

8 Ealing Council, 'Statement – Southall Waterside Health Concerns', January 2021, https://www.ealing.gov.uk/news/article/1950/statement_-_southall_waterside_health_concerns.

9 The Centric Lab, 'About Us', https://www.thecentriclab.com/about-us.

10 Berkeley Group, 'The Green Quarter', https://www.berkeleygroup.co.uk/developments/london/southall/the-green-quarter.

11 Berkeley Group, 'The Green Quarter Community Hub', https://www.thegreenquartercommunity.co.uk/.

12 Sammy Gecsoyler, 'Scientists to Examine Health Fears at West London Luxury Development', *The Guardian*, 27 April 2023, https://www.theguardian.com/business/2023/apr/27/scientists-to-investigate-health-fears-southall-west-london-development-green-quarter.

13 St George, Community Social Impact Report, 2023–24, https://thegreenquartercommunity.co.uk/wp-content/uploads/sites/311/2024/05/StGeorge_AnnualSocialImpactReport.pdf.

14 Gecsoyler, 'Scientists to Examine Health Fears', 2023.

2.6 No safe level of air pollution

1 Friends of the Earth, 'Clean Air: Air Pollution – Why We Need Cleaner Air', https://www.friendsoftheearth.uk/clean-air/.

2 Fiona Harvey, 'Campaigners Call for Unlimited "Climate Card" UK Rail Pass', *The Guardian*, 19 September 2024, https://www.theguardian.com/uk-news/2024/sep/19/campaigners-call-for-unlimited-climate-card-uk-rail-pass.

3 Ibid.

Part Two: Borders

3. Prisons

3.1 Policing each other

1 Largely inspired by the collective Abolitionist Futures.
2 Sarah Lamble, 'Practising Everyday Abolition', *Abolitionist Futures,* https://abolitionistfutures.com/latest-news/practising-everyday-abolition.
3 Odochi Ibe, 'Playing the Game of Respectability Politics, But At What Cost?', 15 February 2022, *Verywell Mind*, https://www.verywellmind.com/playing-the-game-of-respectability-politics-5215862.
4 Prentis Hemphill, 'S2E12: Harm, Punishment, and Abolition with Mariame Kaba', *Finding Our Way* (podcast), https://www.findingourwaypodcast.com/individual-episodes/s2e12.
5 Hemphill, 'S2E12'.
6 Ibid.

3.2 Prison ecology – Climate change from the 'inside'

1 Amber Berg, 'An Introduction to Prison Ecology', *Live Ideas Journal*, November 26, 2019, https://liveideasjournal.com/2019/11/26/an-introduction-to-prison-ecology/.
2 Ryan Sabalow, Dale Kasler and Wes Venteicher, 'Toxic Water in California Prisons: Sickening Inmates and Costing Taxpayers Millions', *The Sacramento Bee*, 3 May 2019, https://www.sacbee.com/news/politics-government/capitol-alert/article229294374.html.
3 Ibid.
4 Michael Waters, 'How Prisons Are Poisoning Their Inmates', *The Outline*, 23 July 2018, https://theoutline.com/post/5410/toxic-prisons-fayette-tacoma-contaminated.

5 Abolitionist Law Center, 'SCI Fayette Coal Ash Investigation', https://abolitionistlawcenter.org/our-work/cases/sci-fayette-coal-ash-investigation/.

6 Kimberly M. S. Cartier, 'America's Toxic Prisons', *The Ecologist*, 13 November 2020, https://theecologist.org/2020/nov/13/environmental-injustice-mass-incarceration-and-systemic-racism-us.

7 Prison Reform Trust, 'Prison: The Facts 2023', https://prisonreformtrust.org.uk/wp-content/uploads/2023/06/prison_the_facts_2023.pdf.

8 Hazel Shearing and Sam Hancock, 'Schools with Dangerous Concrete Race to Replan Start of Term', *BBC News*, 1 September 2023, https://www.bbc.co.uk/news/education-66673971.

9 Ell Folan, 'The UK Has Its Own Mass Incarceration Crisis', *Novara Media*, 2 January 2023, https://novaramedia.com/2023/01/02/the-uk-has-its-own-mass-incarceration-crisis/https://novaramedia.com/2023/01/02/the-uk-has-its-own-mass-incarceration-crisis/.

10 Patrick Daly, 'Jails in Britain Are So Full That Prisoners Are Having to Be Let Out Early. This Criminal Justice Expert Explains What Is Happening', 19 July 2024, https://news.northeastern.edu/2024/07/19/uk-prison-overcrowding/#:~:text=According.

11 Nuray Bulbul, 'How Overcrowded Are the UK's Prisons?', *The Standard*, 13 October, 2023, https://www.standard.co.uk/news/uk/overcrowded-uk-prisons-problem-how-many-b1113284.html.

12 Michael Savage, 'Three-Quarters of Prisons in England and Wales in Appalling Conditions as Overcrowding Fear Grow', *The Guardian*, 5 August 2023, https://www.theguardian.com/society/2023/aug/05/three-quarters-of-prisons-in-england-and-wales-in-appalling-conditions-as-overcrowding-fears-grow.

13 Ibid.

14 Ibid.

15 UK Government, 'Safety in Custody Statistics: England and Wales–Deaths in Prison Custody to June 2023, Assaults and Self-Harm to March 2023', July 2023, https://www.gov.uk/government/statistics/safety-in-custody-quarterly-update-to-march-2023/safety-in-custody-statistics-england-and-wales-deaths-in-prison-custody-to-june-2023-assaults-and-self-harm-to-march-2023.

16 Chantal Edge et al., 'COVID-19 and the Prison Population' (Working Paper) (London: Health Foundation, 2021), 8, https://www.health.org.uk/reports-and-analysis/working-papers/covid-19-and-the-prison-population.

17 Rebecca Tidy, 'Inside Britain's Prisons During the Deadly Heatwave', *Huck Magazine*, 22 July 2022, https://www.huckmag.com/article/inside-britains-prisons-during-the-uk-heatwave.

18 Dominique Moran, Matt Houlbrook and Yvonne Jewkes, 'The Persistence of the Victorian Prison: Alteration, Inhabitation, Obsolescence, and Affirmative Design', *Space and Culture* 25, no. 3 (2022): 364–78, https://doi.org/10.1177/12063312211057036.

19 Ibid.

20 Olenka Frenkiel, 'Prisoners of Katrina', *BBC News*, 10 August 2006, http://news.bbc.co.uk/1/hi/programmes/this_world/5241988.stm.

21 Schuyler Mitchell, 'Hurricane-Struck North Carolina Prisoners Were Locked in Cells with Their Own Feces for Nearly a Week', *The Intercept*, 4 October 2024, https://theintercept.com/2024/10/04/hurricane-helene-north-carolina-mountain-view-prison/.

22 Mansa Musa, 'North Carolina Failed to Evacuate Prisoners During Hurricane Helene', 28 October 2024, https://therealnews.com/north-carolina-failed-to-evacuate-prisoners-during-hurricane-helene.

23 Tidy, 'Inside Britain's Prisons During the Deadly Heatwave'.

24 Ibid.

25 Alyssa Rinaldi, 'Why Was Rikers Island Left Out of the Evacuation Plan for Hurricane Sandy?', *NYU Local*, 5 November 2012,

https://nyulocal.com/why-was-rikers-island-left-out-of-the-evacuation-plan-for-hurricane-sandy-146a9e162c4b.

26 Raven Rakia, 'A Sinking Jail: The Environmental Disaster That Is Rikers Island', *Grist*, 15 March 2016, https://grist.org/justice/a-sinking-jail-the-environmental-disaster-that-is-rikers-island/.

27 Caroline Bankoff, 'Mentally Ill Inmate "Baked to Death" on Rikers Island', *Intelligencer*, 19 March 2014, https://nymag.com/intelligencer/2014/03/inmate-baked-to-death-on-rikers-island.html.

28 Ibid.

29 Rowena Mason, 'Keir Starmer Hits Out at Prison System "Mess" Caused by Tories', *The Guardian*, 6 July 2024, https://www.theguardian.com/politics/article/2024/jul/06/keir-starmer-tribal-politics-four-uk-nations.

30 Full Fact, 'Crime Statistics Explained – Is It Rising or Falling?', *Full Fact*, https://fullfact.org/crime/how-crime-stats-calculated/.

3.3 Being wrong – A social condition

1 Various, 'When You're Black You're Never Really Lonely', *TikTok*, www.tiktok.com/discover/when-youre-black-never-really-lonely?lang=en.

2 Nicholas Turner, 'American History, Race, and Prison', in *Reimagining Prison Web Report*, Vera Institute of Justice, https://www.vera.org/reimagining-prison-web-report/american-history-race-and-prison.

3 Ibid.

4 Ell Folan, 'The UK Has Its Own Mass Incarceration Crisis', *Novara Media*, 2 January 2023, https://novaramedia.com/2023/01/02/the-uk-has-its-own-mass-incarceration-crisis/.

5 J. M. Moore, 'The "New Punitiveness" in the Context of British Imperial History', *CJM 101: #BlackLivesMatter*, Centre for Crime and Justice Studies, 23 September 2015, https://www.

crimeandjustice.org.uk/publications/cjm/article/%E2%80%98new-punitiveness-context-british-imperial-history.

6 Ibid.

7 Turner, 'American History, Race, and Prison'.

8 Critical Resistance, 'The Prison Industrial Complex', https://criticalresistance.org/mission-vision/not-so-common-language/.

9 Sam McCann, 'From Fighting Wildfires to Digging Graves, Incarcerated Workers Face Danger on the Job', *Vera*, 26 July 2023, https://www.vera.org/news/from-fighting-wildfires-to-digging-graves-incarcerated-workers-face-danger-on-the-job.

10 Ibid.

11 Abby Vesoulis, 'Inmates Fighting California Wildfires Are More Likely to Get Hurt, Records Show', *Time*, 16 November 2018, https://time.com/5457637/inmate-firefighters-injuries-death/.

12 Sam Levin, '"Terrifying and Dystopian": The Dark Realities of the Supreme Court's Homelessness Decision', *The Guardian*, 29 June 2024, https://www.theguardian.com/society/article/2024/jun/29/law-professor-homeless-rights-supreme-court-ruling.

13 Joebi-Wan Kenobi, 'Prison Labor and the 13th', *TikTok*, 13 November 2024, https://vm.tiktok.com/ZGd6d1Tj6/.

14 Rhian Lubin and Alicja Hagopian, 'Trump's Immigration Crackdown Mapped: Where Are Majority of ICE Detainees Being Held in the U.S.', *The Independent*, 4 March 2025, https://www.independent.co.uk/news/world/americas/us-politics/trump-immigration-ice-detainees-states-map-b2708881.html.

15 Claire Burrows, 'The Prison Industrial Complex, Racism and Immigration', *Centre for Crime and Justice Studies*, 5 August 2020, https://www.crimeandjustice.org.uk/resources/prison-industrial-complex-racism-and-immigration.

16 JENGbA (Joint Enterprise Not Guilty by Association), 'The Joint Enterprise Law', http://jointenterprise.co/TheLaw.html.

17 Kids of Colour (@KidsOfColourHQ), 'An Update on a Manchester Based Trial Against 10 of Our City's Young People', *X (formerly Twitter)*, 6 May 2022, https://twitter.com/KidsOfColourHQ/status/1522852630988480512/photo/3.

18 UK Parliament, '£4 Billion of Unusable PPE Bought in First Year of Pandemic Will Be Burnt "to Generate Power"', *Public Accounts Committee*, 10 June 2022, https://committees.parliament.uk/committee/127/public-accounts-committee/news/171306/4-billion-of-unusable-ppe-bought-in-first-year-of-pandemic-will-be-burnt-to-generate-power/.

19 British Medical Association, 'COVID-19: What the BMA Is Doing', https://www.bma.org.uk/advice-and-support/covid-19, and British Medical Association, 'COVID-19: The Impact of the Pandemic on the Medical Profession', 18 September 2024, https://www.bma.org.uk/advice-and-support/covid-19/what-the-bma-is-doing/covid-19-the-impact-of-the-pandemic-on-the-medical-profession.

20 Patrick Butler, 'Over 330,000 Excess Deaths in Great Britain Linked to Austerity, Finds Study', *The Guardian*, 5 October 2022, https://www.theguardian.com/business/2022/oct/05/over-330000-excess-deaths-in-great-britain-linked-to-austerity-finds-study.

21 Rachel Hall, 'More Than 1,500 UK Police Officers Accused of Violence Against Women in Six Months', *The Guardian*, 14 March 2023, https://www.theguardian.com/uk-news/2023/mar/14/more-than-1500-uk-police-officers-accused-of-violence-against-women-in-six-months.

3.4 Good trouble – The dilemma of direct action

1 Andrew Boyd and Dave Oswald Mitchell, *Beautiful Trouble: A Toolbox for Revolution* (New York: OR Books, 2012), 182.

2 BBC News, 'Child Q: School Apologises for Strip-Search of Black Schoolgirl', 25 March 2022, https://www.bbc.co.uk/news/uk-england-london-60873858.

3 'Iranian Revolution', *Wikipedia*, last modified 17 April 2024, https://en.wikipedia.org/wiki/Iranian_revolution.

4 'About', *Unis Resist Border Controls*, https://www.unisresistborder controls.org.uk/about/.

5 Leah Lakshmi Piepzna-Samarasinha, *The Future Is Disabled: Prophecies, Love Notes and Mourning Songs*, unabridged audiobook (Old Saybrook, CT: Tantor Audio, 2022), Audible, at 10 min. 51 sec.

6 Ibid., at 8 min. 56 sec.

7 Erica Meltzer, 'Activist Carrie Ann Lucas Told Denver Police to Google How to Use Her Wheelchair; Now She's Charged with Interference', *Denverite*, 30 June 2017, https://denverite.com/2017/06/30/refusing-tell-officers-operate-wheelchair-activist-carrie-ann-lucas-charged-interference/.

8 Piepzna-Samarasinha, *The Future Is Disabled*.

9 Damon Rose, 'The Wheelchair Warriors: Their Rebellious Protests to Change the Law', *BBC News*, https://www.bbc.co.uk/news/extra/8rvpt6bclh/wheelchair-warriors-disability-discrimination-act.

3.5 Prison abolition and the shaky practice of forgiveness

1 'Unity through Diversity – Abolition as Healing', *Highline College Library*, January 2025. https://library.highline.edu/abolition.

2 Ephesians 4:32. See: https://www.biblegateway.com/verse/en/Ephesians%204%3A32.

3 Layli Long Soldier, 'From WHEREAS', *Bomb*, 15 March 2017, https://bombmagazine.org/articles/2017/03/15/from-whereas/.

4 Rob Capriccioso, 'A Sorry Saga: Obama Signs Native American Apology Resolution; Fails to Draw Attention to It,' *Indian Law Resource Center*, Jan 13, 2010, https://indianlaw.org/node/529.

5 David Mercer, 'Metropolitan Police Is "Institutionally Racist, Sexist and Homophobic" and May Have More Officers Like Couzens and Carrick, Review Finds', *Sky News*, 21 March 2023, https://news.sky.com/story/metropolitan-police-is-institutionally-racist-sexist-and-homophobic-and-may-have-more-officers-like-couzens-and-carrick-review-finds-12838717.

6 His Majesty's Inspectorate of Constabulary and Fire & Rescue Services, 'Our latest inspection of @metpoliceuk has been published today' (Tweet regarding HMICFRS status), 9 March 2025, https://x.com/HMICFRS/status/1823978159575838959.

7 Rachel Herzing, 'Abolition Is Practical', *Inquest*, 11 July 2023, https://inquest.org/abolition-is-practical/.

8 Brea Baker, 'Why I Became an Abolitionist', *Harper's Bazaar*, https://www.harpersbazaar.com/culture/politics/a34473938/why-i-became-an-abolitionist/.

9 Emmaline Soken-Huberty, '5 Examples of Restorative Justice', *Global Peace Careers*, https://globalpeacecareers.com/magazine/examples-of-restorative-justice/.

10 Mia Mingus, 'Transformative Justice: A Brief Description', *Transform Harm*, 11 January 2019, https://transformharm.org/tj_resource/transformative-justice-a-brief-description/.

11 Ali Parker, 'Virtually All Rape Victims Are Denied Justice: Here Is the Roadmap to Failure', *Saunders Law*, 3 February 2023, https://www.saunders.co.uk/news/virtually-all-rape-victims-are-denied-justice-here-is-the-roadmap-to-failure/.

12 Baroness Casey of Blackstock, 'Baroness Casey Review: Final Report – An Independent Review into the Standards of Behaviour and Internal Culture of the Metropolitan Police Service',

Metropolitan Police Service, https://www.met.police.uk/SysSiteAssets/media/downloads/met/about-us/baroness-casey-review/update-march-2023/baroness-casey-review-march-2023a.pdf.

4. Disability

1 Talila A. Lewis, 'Working Definition of Ableism (January 2022 Update)', January 2022, https://www.talilalewis.com/blog/working-definition-of-ableism-january-2022-update.

4.1 Bringing disability to the centre

1 Andre Damon, 'Dr. Anthony Fauci Says Many Will "Fall by the Wayside" in New COVID-19 Surge', *World Socialist Web Site*, 31 August 2023, https://www.wsws.org/en/articles/2023/09/01/gqnt-so1.html.

2 United States Holocaust Memorial Museum, 'Eugenics', https://encyclopedia.ushmm.org/content/en/article/eugenics.

3 'The Holocaust', *The National WWII Museum*, 8 May 2024, https://www.nationalww2museum.org/about-us/mission-vision-values.

4 Amanda Tink, 'Disabled People Were Holocaust Victims, Too: They Were Excluded from German Society and Murdered by Nazi Programs', *The Conversation*, 26 January 2023, https://theconversation.com/disabled-people-were-holocaust-victims-too-they-were-excluded-from-german-society-and-murdered-by-nazi-programs-198298.

5 Paola Alonso, 'Autonomy Revoked: The Forced Sterilization of Women of Color in 20th Century America', https://twu.edu/media/documents/history-government/Autonomy-Revoked--The-Forced-Sterilization-of-Women-of-Color-in-20th-Century-America.pdf.

6 Edward Helmore, 'Gen Z Will Be Last Generation with White Majority in US, Study Finds', *The Guardian*, 8 August 2023, https://www.theguardian.com/us-news/2023/aug/08/gen-z-americans-white-majority-study.

7 European Disability Forum, 'EDF Report – September 2022: Forced Sterilisation of Persons with Disabilities in the European Union', https://www.edf-feph.org/content/uploads/2022/09/EDF_FS_0909-accessible.pdf.

8 Maeve Cullinan, 'Inside California's Secret Sterilisation Programme – and Its Antivax Legacy', *The Telegraph*, 16 January 2024, https://www.telegraph.co.uk/global-health/women-and-girls/california-sterilisation-scandal-compensation-hysterectomy/.

9 James Tapper, 'Fury at "Do Not Resuscitate" Notices Given to Covid Patients with Learning Disabilities', *The Guardian*, 13 February 2021, https://www.theguardian.com/world/2021/feb/13/new-do-not-resuscitate-orders-imposed-on-covid-19-patients-with-learning-difficulties.

10 Ibid.

11 KFF, 'Global COVID-19 Tracker', *KFF*, 24 March 2025, https://www.kff.org/coronavirus-covid-19/issue-brief/global-covid-19-tracker/.

12 Daniel DeWitt, 'Richard Dawkins' Dangerous Tweet Re: Eugenics', *The Olatte*, 18 February 2020, https://www.theolatte.com/2020/02/richard-dawkins-dangerous-tweet-re-eugenics/.

13 Schalk, *Black Disability Politics*, 144.

14 Ibid., 9.

15 World Health Organization, 'Disability – Key Facts', 7 March 2023, https://www.who.int/news-room/fact-sheets/detail/disability-and-health.

16 Susan Sontag, *Illness as Metaphor & AIDS and Its Metaphors* (London: Penguin Modern Classics, 2009), 6.

17 Sam Levin, '"A Talented, Goofy Kid": Family of Ryan Gainer, Autistic Teen Killed by Police, Speak Out', *The Guardian*, 21 March 2024, https://www.theguardian.com/us-news/2024/mar/21/ryan-gainer-autistic-teen-police-killing-california.

18 Mental Health Foundation, 'Black, Asian and Minority Ethnic (BAME) Communities', *Mental Health Awareness Week*, https://www.mentalhealth.org.uk/explore-mental-health/a-z-topics/black-asian-and-minority-ethnic-bame-communities.

4.2 Black disability and climate change

1 Imani Barbarin (@Imani_Barbarin), 'Ableism Is What Makes All Other "Isms" Effective ...', *X* (formerly Twitter), 14 August 2023, https://twitter.com/Imani_Barbarin/status/1691149414868332558.

2 Equal Justice Initiative, 'Medical Exploitation of Black women', *Equal Justice Initiative*, 29 August 2019, https://eji.org/news/history-racial-injustice-medical-exploitation-of-black-women/.

3 Daniel J. Weiss, 'Fossil Fuel Industries Kill and Injure an Awful Lot of Their Workers', *Grist*, 20 April 2011, https://grist.org/fossil-fuels/2011-04-19-fossil-fuel-industries-kill-injure-workers/.

4 Sara Sneath, 'Fossil Fuel Workers Are Dying Inhaling Gases – Despite US Warnings to Big Oil', *The Guardian*, 13 July 2023, https://www.theguardian.com/us-news/2023/jul/13/fossil-fuel-deaths-inhaling-gas.

5 Patrick Vernon, 'The Windrush Scandal Was Traumatic. Survivors Need Tailored Mental Health Care', *The Guardian*, 9 October 2019, https://www.theguardian.com/society/2019/oct/09/windrush-scandal-survivors-mental-health-care.

6 BBC News, 'Black Mould: How Dangerous Is It in the Home and How Can It Be Treated?', 13 February, https://www.bbc.co.uk/news/uk-63642856.

7 Lucie Heath, 'More Than 100 Buildings with Grenfell-Style Cladding Yet to Complete Work Nearly Five Years after Tragedy', *Inside Housing*, 19 May 2022, https://www.insidehousing.co.uk/news/more-than-100-buildings-with-grenfell-style-cladding-yet-to-complete-work-nearly-five-years-after-tragedy-75689.

8 Yasmeen Serhan, 'Was London's Grenfell Tower Fire Preventable?', *The Atlantic*, 19 June 2017, https://www.theatlantic.com/international/archive/2017/06/was-londons-grenfell-tower-fire-preventable/530589/.

9 Rooted in Rights, 'Audio Described: The Right to Be Rescued', *YouTube* video, uploaded 2017, https://www.youtube.com/watch?v=FIFEvqafklk.

10 Ibid.

11 Jordan Melograna, 'People with Disabilities Left Behind During Katrina Tell Their Stories', *HuffPost*, 27 August 2015, https://www.huffpost.com/entry/people-with-disabilities_b_8045700.

12 Bill Quigley, 'Six Months after Katrina: Who Was Left Behind', Common Dreams via Global Action on Aging, 21 February 2006, https://globalag.igc.org/armedconflict/countryreports/americas/sixkatrina.htm.

13 Max Airborne 'Fat Roots: Emmett Everett', 5 December, 2019, *Fat Rose*, https://fatrose.org/2019/12/05/remember-emmett-everett/.

14 Ibid.

15 Sankeerth Achalla, 'The Impact of Heatwaves on People with Disabilities', *Eztia*, https://www.eztiahealth.com/blog/impact-of-heatwaves-on-people-with-disabilities.

16 Lauren Geall, 'Can Taking Antidepressants Put You at Greater Risk of Heat Stroke and Dehydration?', *Stylist*, July 2024, https://www.stylist.co.uk/health/mental-health/antidepressants-heat-intolerance-sweating/684900.

17 Joshua Nelken-Zitser and Grace Eliza Goodwin, 'Price-Gouging Complaints about the Cost of Fuel, Water, and Hotels Are

Surging in States Hit by Hurricane Helene', *Business Insider*, 2 October, 2024, https://www.businessinsider.com/price-gouging-complaints-surge-hurricane-helene-states-fuel-water-hotels-2024-10.

18 Jake Johnson, 'Corporate Price Gouging Amid Hurricanes Prompts Calls for Federal Action', *Truthout*, 9 October 2024, https://truthout.org/articles/corporate-price-gouging-amid-hurricanes-prompts-calls-for-federal-action/.

19 European Environment Agency, 'Europe Is Not Prepared for Rapidly Growing Climate Risks', 10 March 2024, https://www.eea.europa.eu/en/newsroom/news/europe-is-not-prepared-for.

4.3 The miseducation of Covid – A primer

1 James Baldwin, 'The Precarious Vogue of Ingmar Bergman', *Esquire*, 1 April 1960, https://classic.esquire.com/article/1960/4/1/the-precarious-vogue-of-ingmar-bergman.

2 WHO Team, Emergency Response (WRE), 'COVID-19 Epidemiological Update – 19 January 2024', edition 163, *World Health Organization*, 19 January 2024, https://www.who.int/publications/m/item/covid-19-epidemiological-update---19-january-2024.

3 Edouard Mathieu et al., 'Coronavirus (COVID-19) Deaths', *Our World in Data*, 2020, https://ourworldindata.org/covid-deaths.

4 'The Pandemic's True Death Toll', *The Economist*, 25 October 2022, https://www.economist.com/graphic-detail/coronavirus-excess-deaths-estimates.

5 'Coronavirus Death Toll', *Worldometer*, last updated 13 April 2024, https://www.worldometers.info/coronavirus/coronavirus-death-toll/#google_vignette.

6 Bastian Herre, Lucas Rodés-Guirao and Max Roser, 'War and Peace', *Our World in Data*, https://ourworldindata.org/war-and-peace#:~:text=But%%20per%20cent%20the%%20

per%20cent20deaths%%20per%20cent20in%%20per%20
cent20the,than%%20per%20cent2021%%20per%20cent-
20million%%20per%20cent20combatants%%20per%20
cent20died.

7 'Epidemiology of HIV/AIDS', *Wikipedia*, https://en.wikipedia.
org/wiki/Epidemiology_of_HIV/AIDS.

8 Dani Blum, 'This May Be the Most Overlooked Covid Symptom', *The New York Times*, 5 July 2024, https://www.nytimes.
com/2024/07/05/well/covid-symptoms-stomach-pain-
diarrhea.html.

9 'Biosafety Level Guidance for COVID-19 Research', Consteril,
https://consteril.com/biosafety-level-guidance-covid-19-research.

10 Louwers Media Group, 'Professional Air Purifiers Can Lead
to a 90% Lower Chance of Indirect Covid-19 Infection', 14
October 2022, https://www.installatieenbouw.be/artikel/
professionele-luchtreinigers-kunnen-leiden-tot-90-kleinere-kans-
op-indirecte-covidCovid-19-besmetting/.

11 Marianne Cooper and Maxim Voronov, 'We've Hit Peak Denial.
Here's Why We Can't Turn Away from Reality', *Scientific
American*, 18 June 2024, https://www.scientificamerican.com/
article/weve-hit-peak-denial-heres-why-we-cant-turn-away-
from-reality/.

12 National Academies of Sciences, Engineering, and Medicine,
'Long-Term Health Effects of COVID-19: Disability and Function
Following SARS-CoV-2 Infection' Washington, DC: The National
Academies Press, 2024, https://nap.nationalacademies.org/
catalog/27756/long-term-health-effects-of-covid-19-disability-
and-function.

13 Blum, 'This May Be the Most Overlooked Covid Symptom'.

14 *Financial Times*, '"Immunity Debt" Is a Misguided and Dangerous
Concept', https://www.ft.com/content/0640004d-cc15-481e-90ce-
572328305798.

15 Rachel Clun, 'Multiple Hospitals Declare Critical Incidents over Soaring Flu Cases as A&E Patients Face 50 Hour Waits', *The Independent*, Wednesday 8 January 2025, https://www.independent.co.uk/news/health/royal-liverpool-hospital-critical-incident-flu-cold-winter-b2675013.html.

16 Catherine Demoliou et al., 'SARS-CoV-2 and HIV-1: So Different Yet So Alike. Immune Response at the Cellular and Molecular Level', *International Journal of Medical Sciences* 19, no. 12 (2022): 1787–1795, published 3 October 2022, https://www.ncbi.nlm.nih.gov/pmc/articles/PMC9608044/.

17 Jennifer Henderson, 'Most COVID Transmission Is Still Asymptomatic', *ABC News*, 11 May 2022, https://abcnews.go.com/Health/covidCovid-transmission-asymptomatic/story?id=84599810.

18 Global Center for Health Security, 'Long COVID Persists as a Mass Disabling Event', *University of Nebraska Medical Center*, 25 July 2023, https://www.unmc.edu/healthsecurity/transmission/2023/07/25/long-covid-persists-as-a-mass-disabling-event/.

19 Office for National Statistics (ONS), 'Prevalence of Ongoing Symptoms Following Coronavirus (COVID-19) Infection in the UK', *Statistical Bulletins*, 30 March 2023, https://www.ons.gov.uk/peoplepopulationandcommunity/healthandsocialcare/conditionsanddiseases/bulletins/prevalenceofongoingsymptomsfollowingcoronaviruscovidCovid19infectionintheuk/30march2023.

20 Ziyad Al-Aly et al., 'Long COVID Science, Research and Policy', *Nature Medicine* 30 (2024): 2148–2164, https://www.nature.com/articles/s41591-024-03173-6.

21 US Centers for Disease Control and Prevention (CDC), 'COVID-19 Signs and Symptoms of Long COVID', https://www.cdc.gov/covidCovid/long-term-effects/long-covidCovid-signs-symptoms.html.

22 Suchitra Rao et al. 'Postacute Sequelae of SARS-CoV-2 in Children', *Pediatrics*, 153, no. 3 (2024), e2023062570. https://doi.org/10.1542/peds.2023-062570.

23 Hayley Gleeson, 'Too Many Children with Long COVID Are Suffering in Silence. Their Greatest Challenge? The Myth That the Virus Is "Harmless" for Kids', *ABC News (Australia)*, 15 June 2024, https://www.abc.net.au/news/2024-06-16/children-with-long-covidCovid-dismissed-doctors-myth-virus-harmless/103959078.

24 Rao, et al., 'Postacute Sequelae of SARS-CoV-2 in Children'.

25 @lizzie_traveler_public Instagram, '#covid19 is a neurotrophic vascular illness that causes cumulative, potentially permanent, damage to the brain and shouldn't be taken lightly'. https://www.instagram.com/reel/DDjDPpcxJPS/?igsh=OWU1OWpoMXRleXYy.

26 Carlos Oliveira, 'Long COVID in Children: What Parents Need to Know', *UNICEF Parenting*, https://www.unicef.org/parenting/health/long-COVID-children.

27 Hannah Cromarty, 'Coronavirus: Support for Rough Sleepers (England)', *House of Commons Library*, 12 October 2021, https://commonslibrary.parliament.uk/research-briefings/cbp-9057/.

28 Parents' Coalition of Montgomery County, Maryland, 'Biden Brings His Own Ventilation to @mcps High School #COVIDisAirborne #CO2', 26 August 2022, https://parentscoalitionmc.blogspot.com/2022/08/biden-brings-his-own-ventilation-to.html.

29 Zeke Miller, 'White House Lifting Its COVID-19 Testing Rule for People Around Biden, Ending a Pandemic Vestige', *AP News*, 4 March 2024, https://apnews.com/article/white-house-covidCovid-testing-biden-masks-7a68a0588700d1c3cd0b6f800be805a9.

30 Claire Gilbody Dickerson, 'Joe Biden Contracts COVID as Pressure for Him to Exit Presidential Race Mounts', *Sky News*,

18 July 2024, https://news.sky.com/story/joe-biden-pulls-out-of-speech-after-contracting-covidCovid-13179594.

31 @CounsellingSam, 'King Charles has HEPA filters in Buckingham Palace. Why can't we?', *X* (formerly Twitter), 24 February 2024, https://x.com/CounsellingSam/status/1760627437912956976.

32 Eastern Economic Forum, 'The Procedure for Passing PCR Tests for EEF-2024', 30 August 2024, https://forumvostok.ru/en/news/the-procedure-for-passing-pcr-tests-for-eef-2024/.

33 'New Type of Ultraviolet Light Makes Indoor Air as Safe as Outdoors', *Columbia University Irving Medical Center*, 25 March 2022, https://www.cuimc.columbia.edu/news/new-type-ultraviolet-light-makes-indoor-air-safe-outdoors.

34 @_CatintheHat, 'Of course, it's not just staff affected by high rates of Covid transmission in hospitals', Tweet posted on 10 April 2025, X (formerly Twitter), https://x.com/_CatintheHat/status/1861062962913366325.

35 @_CatintheHat, 'However, NHS staff were being given a very DIFFERENT message at that time', Tweet posted on 10 April 2025. X (formerly Twitter). https://x.com/_CatintheHat/status/1860782263367020711.

36 And so, if you find yourself questioning why I still speak about Covid-19 in this book, come back to this section.

4.4 Covid – The magnifier

1 Steven W. Thrasher, *The Viral Underclass: The Human Toll When Inequality and Disease Collide* (New York: Celadon Books, 2022), 19.

2 Ibid., 43–44.

3 Heather Pringle, 'How Europeans Brought Sickness to the New World', *Science*, 4 June, 2015, https://www.science.org/content/article/how-europeans-brought-sickness-new-world-rev2.

4 Mark Twain, 'History Doesn't Repeat Itself, but It Does Rhyme', *Goodreads*, https://www.goodreads.com/quotes/5382-history-doesn-t-repeat-itself-but-it-does-rhyme.

5 Oxfam International, 'Reaction to News that EU Discarded at Least 215 Million Doses of COVID-19 Vaccine', 18 December 2023, https://www.oxfam.org/en/press-releases/reaction-news-eu-discarded-least-215-million-doses-covid-19-vaccine.

6 Luisa Frallonardo et al., 'Incidence and Burden of Long COVID in Africa: A Systematic Review and Meta-Analysis', *Scientific Reports* 13 (2023): 21482, https://doi.org/10.1038/s41598-023-48258-3.

7 'UK PM's Former Adviser Confirms Johnson Said "Let the Bodies Pile High"', *Reuters*, May 26, 2021, https://www.reuters.com/world/uk/uk-pms-former-adviser-confirms-johnson-said-let-bodies-pile-high-2021-05-26/.

8 Robert Booth and Caelainn Barr, 'Black People Four Times More Likely to Die from Covid-19, ONS Finds', *The Guardian*, 7 May 2020, https://www.theguardian.com/world/2020/may/07/black-people-four-times-more-likely-to-die-from-covid-19-ons-finds.

9 Adrian Florido, 'White People Feared COVID Less after Learning Other Races Were Hit Hardest, Data Show', *NPR*, 4 April 2022, https://www.npr.org/2022/04/04/1090919953/white-people-feared-covid-less-after-learning-other-races-were-hit-hardest-data-.

10 'Denying COVID-19 Vaccines to Palestinians Exposes Israel's Institutionalized Discrimination', *Amnesty International*, 6 January 2021, https://www.amnesty.org/en/latest/press-release/2021/01/denying-covid19-vaccines-to-palestinians-exposes-israels-institutionalized-discrimination/.

11 Sontag, *Illness as Metaphor*, 130.

12 Oliver Milman, '"Potentially Devastating": Climate Crisis May Fuel Future Pandemics', *The Guardian*, 28 April 2022, https://www.theguardian.com/environment/2022/apr/28/climate-crisis-future-pandemics-zoonotic-spillover.

4.5 Saying what you need

1 Alison Kafer, *Feminist, Queer, Crip*, Indiana University Press, 2013, Notes chapter, at 26 min. 30 sec.
2 Ibid., at 42 min. 50 sec.
3 Ibid., 32 min. 55 sec.
4 Schalk, *Black Disability Politics*, 9.
5 Johanna Hedva, *Sick Woman Theory* (2020), https://www.kunstverein-hildesheim.de/assets/bilder/caring-structures-ausstellung-digital/Johanna-Hedva/cb6ec5c75f/AUSSTELLUNG_1110_Hedva_SWT_e.pdf.
6 Sara Hendren, 'Our Bodies, Aliveness, and the Built World', *On Being with Krista Tippett*, 16 November 2023, https://onbeing.org/programs/sara-hendren-our-bodies-aliveness-and-the-built-world/#transcript.
7 Ibid.
8 Hedva, *Sick Woman Theory*, 8.

4.6 Our disabled futures

1 Leah Lakshmi Piepzna-Samarasinha, '"The Future Is Disabled" Imagines a World Oriented Around Care and Safety', *Teen Vogue*, 4 October 2022, https://www.teenvogue.com/story/future-is-disabled-book.
2 'Planning for Climate Impact Falls Short Once Again', *Committee on Climate Change*, 13 March 2024, https://www.theccc.org.uk/2024/03/13/planning-for-climate-impacts-falls-short-once-again/.

3 Thomas Johnson, 'Government "Falls Short" in Response to Report about Lack of Extreme Heat Resilience', *New Civil Engineer*, 30 April 2024, https://www.newcivilengineer.com/latest/government-falls-short-in-response-to-report-about-lack-of-extreme-heat-resilience-30-04-2024/.

4 'Local Residents Left in the Dark About Dangerous Air Pollution', *UK Parliament*, 26 October 2022, https://committees.parliament.uk/committee/127/public-accounts-committee/news/173865/local-residents-left-in-the-dark-about-dangerous-air-pollution/.

5 People's CDC, 'A CDC Watchdog and Public Health Advocacy and Health Justice Group', https://peoplescdc.org.

6 Julia Doubleday, 'Anti-COVID Groups Distribute Masks and Air Purifiers Faster Than LA Government Amidst Fires', *The Gauntlet*, 16 January 2025, https://www.thegauntlet.news/p/anti-covid-groups-distribute-masks.

7 Allan M. Brandt, 'How AIDS Activists Fought for Patients' Rights', *History.com*, 30 November 2022, https://www.history.com/news/act-up-aids-patient-rights.

8 Ibid.

9 Historyin3 (@historyin3), 'Survival Programs', *TikTok*, https://vm.tiktok.com/ZGewXQVJN/.

10 Huey P. Newton Foundation, *The Black Panther Party: Service to the People Programs*, ed. David Hilliard (University of New Mexico Press, 15 May 2008), 21–22.

11 MAIA, 'Rehearsing Liberation into Being', https://www.maia-group.co/.

5. Migration

5.1 The empire within us

1 Jónína Kirton, 'Reconciliation', *On Being*, 3 May 2021, https://onbeing.org/poetry/reconciliation/.

2 James Chen, 'BRICS: Acronym for Brazil, Russia, India, China, and South Africa', *Investopedia*, 11 August 2024, https://www.investopedia.com/terms/b/brics.asp.

3 Rédaction Africanews with Agencies, 'Mali Drops French as Official Language', *Africanews*, last updated 13 August 2023, https://www.africanews.com/2023/07/26/mali-drops-french-as-official-language/.

4 Cat Wiener and BBC Monitoring, 'Mali Wins $160m in Gold Mining Dispute after Detaining British Businessman', *BBC News*, 18 November 2024, https://www.bbc.co.uk/news/articles/clyg6319d1eo.

5 'Senegal's Sonko Wants to Reassess Ties with France', *Africanews*, March 2024, *YouTube*, https://www.youtube.com/watch?v=5x9Mt2kmrp4.

6 Military Africa, 'Senegal Asks France to Close Military Bases', *Military Africa*, 3 December 2024, https://www.military.africa/2024/12/senegal-asks-france-to-close-military-bases/.

7 Le Monde with AFP, 'Côte d'Ivoire President Says French Forces to Withdraw in January', *Le Monde*, 1 January 2025, https://www.lemonde.fr/en/le-monde-africa/article/2025/01/01/cote-d-ivoire-president-says-french-forces-to-withdraw-in-january_6736618_124.html.

8 Paul Njie, 'Chad Cuts Military Agreement with France', *BBC News*, 29 November 2024, https://www.bbc.co.uk/news/articles/c7042v17kjqo.

9 Nimi Princewill, 'Macron's Claim That Africans Failed to Say "Thank You" for French Military Aid Sparks Outrage', *CNN*, 7 January 2025, https://edition.cnn.com/2025/01/07/africa/macron-africa-comments-spark-outrage-intl/index.html.

10 Oumar Sankare, 'France's Macron "Insulting All Africans," Says Burkina Faso President', *Anadolu Agency*, 14 January 2025, https://www.aa.com.tr/en/africa/frances-macron-insulting-all-africans-says-burkina-faso-president/3450047#.

11 Joseph Atainyang, 'Burkina Faso's President Rejects Loan from IMF, Shuns Trump's Invitation', *Watchman Post*, 22 March 2025, https://watchmanpost.ng/burkina-fasos-president-rejects-loan-from-imf-shuns-trumps-invitation/.

12 Sang N., 'This Is What President Ibrahim Traoré of Burkina Faso Has Achieved in 2 Years', *Malawi Ace*, 7 February 2025, https://malawiace.com/2025/02/07/this-is-what-president-ibrahim-traore-of-burkina-faso-has-achieved-in-2-years/.

13 Arise News, 'President Ibrahim Traoré Prohibits British and French Judicial Wigs', *Arise News*, 14 January 2025, https://www.arise.tv/president-ibrahim-traore-prohibits-british-and-french-judicial-wigs/.

14 Al Jazeera, 'Gaza Is "Most Documented Genocide in History," Says Palestinian UN Rep', *Al Jazeera*, 17 July 2024, https://www.aljazeera.com/program/newsfeed/2024/7/17/gaza-is-most-documented-genocide-in-history-says-palestinian-un-rep.

15 BBC News, 'Barbados Becomes a Republic and Parts Ways with the Queen', *BBC News*, 30 November 2021, https://www.bbc.co.uk/news/world-latin-america-59470843.

16 Anna Fleck, 'Which Nations Want to Cut Ties with the British Monarchy? Jamaica Referendum', *Statista*, 28 September 2023, https://www.statista.com/chart/30904/commonwealth-countries-who-would-vote-to-become-a-republic/.

17 Katy Watson and Daniela Relph, 'Indigenous Australian Senator Defends Heckling King', *BBC News*, 21 October 2024, https://www.bbc.co.uk/news/articles/c79n2or750po.

18 Amelia Hill, 'British Public Support for Monarchy at Historic Low, Poll Reveals', *The Guardian*, 28 April 2023, https://www.theguardian.com/uk-news/2023/apr/28/public-support-monarchy-historic-low-poll-reveals.

19 Cary Aspinwall, 'Some States Are Turning Miscarriages and Stillbirths Into Criminal Cases Against Women', *The Marshall Project*, 31

October 2024, https://www.themarshallproject.org/2024/10/31/stillbirth-oklahoma-arkansas-women-investigated.

20 Hugo Lowell and Rachel Leingang, 'Trump Signs Executive Order to Dismantle US Department of Education', *The Guardian*, 20 March 2025, https://www.theguardian.com/us-news/2025/mar/20/trump-executive-order-education-department.

21 Kriston Capps, 'Trump DEI Purge Hits Affordable Housing Groups', *Bloomberg*, 11 March 2025, https://www.bloomberg.com/news/articles/2025-03-11/doge-cuts-contracts-to-affordable-housing-groups-who-mention-dei-words.

22 Thomas Naadi, 'Stevie on the Wonder of Becoming a Ghanaian Citizen', *BBC News*, 14 May 2024, https://www.bbc.co.uk/news/articles/c4n1137nj290.

23 Emmanuel Akinwotu, 'A New Home for the African Diaspora in Ghana Stirs Tensions', *NPR*, 25 February 2024, https://www.npr.org/2024/02/25/1225192589/a-new-home-for-the-african-diaspora-in-ghana-stirs-tensions.

24 Ibid.

25 Saidiya Hartman, *Lose Your Mother: A Journey Along the Atlantic Slave Route* (London: Serpent's Tail, 2021), 33.

26 Ibid., 100.

5.2 A question of belonging

1 betteraveparfaite, 'Sissako – Extrait du film Bamako', *YouTube*, 2013, https://www.youtube.com/watch?v=lg49b18lTVk.

2 British Red Cross, 'Asylum Seekers: Are They Living on Easy Street?', *British Red Cross*, n.d., https://www.redcross.org.uk/stories/migration-and-displacement/refugees-and-asylum-seekers/asylum-seekers-are-they-living-on-easy-street.

3 Migrant Help, 'Aspen Cards', updated 6 January 2025, https://www.migranthelpuk.org/pages/faqs/category/aspen.

4 Jully, 'Heartbreak Hotels: Hotels Are Breaking People Seeking Asylum', *Refugee Action*, https://www.refugee-action.org.uk/heartbreak-hotels/.

5 Karolina Olczak, 'Migrants' Experiences of the UK Immigration System: Key Insights', 2 December 2024, https://www.freeths.co.uk/insights-events/legal-articles/2024/migrants-experiences-of-the-uk-immigration-system-key-insights/.

6 Rajeev Syal, 'Suella Braverman Claims "Hurricane" of Mass Migration Coming to UK,' *The Guardian*, 3 October 2023, https://www.theguardian.com/politics/2023/oct/03/suella-braverman-claims-hurricane-of-mass-migration-coming-to-uk.

7 Graeme Demianyk, '"Dystopian": Rishi Sunak's New "Stop The Boats" Lectern Prompts Backlash', *HuffPost UK*, 7 March 2023, https://www.huffingtonpost.co.uk/entry/rishi-sunak-lectern-stop-the-boats_uk_640778abe4b0bbbc6b2f3f65.

8 Rajeev Syal, 'Starmer accused of echoing far right with "island of strangers" speech', *Guardian*, 12 May 2025, https://www.theguardian.com/politics/2025/may/12/keir-starmer-defends-plans-to-curb-net-migration.

9 Trending Now, 'Tommy Robinson Arrested in Canada', *YouTube* video, July 2024, https://www.youtube.com/watch?v=tmeRF-QTQss.

10 Christian Dustmann and Tommaso Frattini, *The Fiscal Effects of Immigration to the UK* (London: University College London, 27 November 2013), https://www.ucl.ac.uk/european-institute/news/2013/nov/fiscal-effects-immigration-uk.

11 Anne Morris, 'Immigration & Societal Contributions', *Davidson-Morris*, 12 February 2025, https://www.davidsonmorris.com/immigration-societal-contributions/#:~:text=Statistics%%20per%20cent20Showing%%20per%20cent20the%%20per%20

cent2oImpact%%2oof,the%%2oUK%%2oper%2ocent27s%%2o
per%2ocent2oeconomic%%2ooutput%%2oper%2ocent2oannually.

12 The Migration Observatory, *Net Migration to the UK*, 2 December 2024, https://migrationobservatory.ox.ac.uk/resources/briefings/long-term-international-migration-flows-to-and-from-the-uk/.

13 Ben Brindle and Madeleine Sumption, 'How Will New Salary Thresholds Affect UK Migration?' 6 December 2023, https://ukandeu.ac.uk/how-will-new-salary-thresholds-affect-uk-migration/.

14 Georgina Sturge, Sonja Stiebahl and Cassie Barton, *Asylum Statistics*, Research Briefing (UK Parliament), 4 March 2025, https://commonslibrary.parliament.uk/research-briefings/sn01403/.

15 Christa Rottensteiner and Claire Kumar, 'As UK Public Attitudes Toward Migration Are Increasingly Positive, It's Time for More Balanced and Evidence-Based Narratives', *IOM Blog*, International Organization for Migration, 25 April 2023, https://weblog.iom.int/uk-public-attitudes-toward-migration-are-increasingly-positive-its-time-more-balanced-and-evidence-based-narratives.

5.3 The other side of the border

1 'Maldives', *Wikipedia*, https://en.wikipedia.org/wiki/Maldives.

2 United Nations Environment Programme (UNEP), 'UNEP, Maldives Partner to Address Climate Change and Other Environmental Threats', 30 August 2024, https://www.unep.org/news-and-stories/story/unep-maldives-partner-address-climate-change-and-other-environmental-threats.

3 Lawrence Huang, 'Climate Migration 101: An Explainer', *Migration Policy Institute*, 16 November 2023, https://www.migrationpolicy.org/article/climate-migration-101-explainer.

4 Florence Stuart, 'Fairbourne: The Village That Could Be Lost to the Sea', *Greenpeace*, 13 March 2020, https://www.greenpeace.org.uk/news/fairbourne-village-lost-sea-climate-change/.

5 Greenpeace UK, 'How Will Climate Change Affect the UK?' https://www.greenpeace.org.uk/challenges/climate-change/how-will-climate-change-affect-the-uk/.

6 Georgie Hughes, 'UK Cities Expected to Face Water Shortages by 2040 Revealed', *Environment Journal*, 4 August 2022, https://environmentjournal.online/water/uk-cities-expected-to-face-water-shortages-by-2040-revealed/.

7 Laura Sharman, 'London Flood Map Shows Areas of City at Risk of Being Underwater Within 10 Years', *Evening Standard*, 12 August 2021, https://www.standard.co.uk/news/london/london-flooding-risk-map-areas-underwater-10-years-b950199.html.

8 Jess Warren, 'Climate Change: London Is Underprepared for Extreme Weather', *BBC News*, 17 January 2024, https://www.bbc.co.uk/news/uk-england-london-67993950.

9 'UK Weather: Flooding In London After Heavy Rain', *BBC News*, 25 July 2021, https://www.bbc.co.uk/news/uk-england-london-57963856.

10 Huang, 'Climate Migration 101'.

11 Climate & Migration Coalition, 'Getting Started: Our 20 Minute Introduction to Climate Change and Migration', *Climate Migration Coalition*, https://climatemigration.org.uk/getting-started-climate-migration/.

12 Ibid.

13 Niuone Eliuta, 'Science Says Tuvalu Will Drown Within Decades; the Reality Is Worse', *ReliefWeb*, 15 February 2024, https://reliefweb.int/report/tuvalu/science-says-tuvalu-will-drown-within-decades-reality-worse.

14 Ayesha Tandon, 'In-depth Q&A: How Does Climate Change Drive Human Migration?', *CarbonBrief*, 11 April 2024, https://interactive.carbonbrief.org/climate-migration/index.html.

15 Voskoboynik, 'To Fix the Climate Crisis, We Must Face Up to Our Imperial Past'.

16 Ibid.

17 'Sierra Madre (Philippines)', *Wikipedia*, last modified 21 July 2003, https://en.wikipedia.org/wiki/Sierra_Madre_%28Philippines%29.

18 Climate Change Commission, 'Sierra Madre: Mountain Range for Resilience', 26 September 2024, https://climate.gov.ph/news/934.

19 Nica Glorioso, 'Why Should We Protect the Sierra Madre, the "Backbone of Luzon"?', *Nylon Manila*, 19 November 2024, https://nylonmanila.com/voice/why-should-we-protect-sierra-madre/.

20 Laura Bicker, 'Sierra Madre: Fighting to Save What's Left of a Vital Rainforest', *BBC News*, 5 January 2023, https://www.bbc.co.uk/news/world-asia-64123652.

21 Ibid.

22 Greg Bankoff, 'One Island Too Many: Reappraising the Extent of Deforestation in The Philippines Prior to 1946', *Journal of Historical Geography* 33, no. 2 (April 2007): 314–334, https://www.researchgate.net/publication/248584674_One_island_too_many_reappraising_the_extent_of_deforestation_in_the_Philippines_prior_to_1946.

23 Dorothy Guerrero, 'Colonialism, Climate Change and Climate Reparations', *Global Justice Now*, 4 August 2023, https://www.globaljustice.org.uk/blog/2023/08/colonialism-climate-change-and-climate-reparations/.

5.4 Climate migration – A new reality?

1 Reuters, 'What to Know About the El Salvador Mega-Prison Where Trump Sent Deported Venezuelans', *The Guardian*, 20 March 2025, https://www.theguardian.com/world/2025/mar/20/cecot-el-salvador-venezuela-prison-trump-deportations.

2 NowThis Impact, 'Families of People Sent to a Mega-Prison In El Salvador Are Speaking Out, Saying Their Loved Ones Were

Taken There Despite Having No Gang Affiliation Or Criminal Record', *TikTok* video, https://vm.tiktok.com/ZNd8hxDgq/.

3 Ibid.

4 Reuters, What to Know About the El Salvador Mega-Prison'.

5 Oli Mould, 'Under Flashism, Cruelty Is the Point', *TikTok* video, https://vm.tiktok.com/ZNdLXAMpr/.

6 Institute for Economics & Peace, 'Over One Billion People at Threat of Being Displaced by 2050 Due to Environmental Change, Conflict and Civil Unrest', *Ecological Threat Register*, 27 August 2020, https://www.economicsandpeace.org/wp-content/uploads/2020/09/Ecological-Threat-Register-Press-Release-27.08-FINAL.pdf.

7 Sarah Nash and Caroline Zickgraf, 'Stop Peddling Fear of Climate Migrants', 23 September 2020, *openDemocracy*, https://www.opendemocracy.net/en/stop-peddling-fear-climate-migrants/.

8 Huang, 'Climate Migration 101'.

9 Ibid.

10 Gaia Vince, *Nomad Century: How Climate Migration Will Reshape Our World* (New York: Flatiron Books, 22 August 2023), 10.

11 Laura Paddison, 'Siberia Swelters in Record-Breaking Temperatures Amid its "Worst Heat Wave In History"', 8 June 2023, *CNN*, https://www.cnn.com/2023/06/08/world/siberia-heatwave-record-temperatures-intl-scn/index.html.

12 Leyland Cecco, '"It's Devastating": Summer in Canada's Arctic Region Brings Severe Heatwaves', 8 August 2024, *The Guardian*, https://www.theguardian.com/world/article/2024/aug/08/canada-arctic-region-heat-wave.

13 Vince, *Nomad Century*, 107–109.

14 Johann Harnoss, Tanya Mondal, and Janina Kugel, 'Will a Green Skills Gap of 7 Million Workers Put Climate Goals at Risk?', 14 September 2023, *Boston Consulting Group*, https://www.bcg.com/

publications/2023/will-a-green-skills-gap-put-climate-goals-at-risk.

15 Ibid.

16 Eric Koons, 'Renewable Energy in Nepal: Building a Sustainable Future', 11 April 2024, *Climate Impacts Tracker*, https://www.climateimpactstracker.com/renewable-energy-in-nepal-building-a-sustainable-future/.

17 Rapid Transition Alliance, 'Doing Development Differently: How Kenya Is Rapidly Emerging as Africa's Renewable Energy Superpower', 17 November 2022, https://rapidtransition.org/stories/doing-development-differently-how-kenya-is-rapidly-emerging-as-africas-renewable-energy-superpower/.

18 The Migration Observatory, *Migration and the Health and Care Workforce*, 27 June 2023, https://migrationobservatory.ox.ac.uk/resources/briefings/migration-and-the-health-and-care-workforce/.

19 'Our Ageing Population | The State of Ageing 2023–24', *The Centre for Ageing Better*, https://ageing-better.org.uk/our-ageing-population-state-ageing-2023-4.

20 Sabelo J. Ndlovu-Gatsheni, '"Moral Evil, Economic Good": Whitewashing the Sins of Colonialism', 26 February 2021, *Al Jazeera*, https://www.aljazeera.com/opinions/2021/2/26/colonialism-in-africa-empire-was-not-ethical.

21 Trevor Getz, 'African Resistance to Colonialism', *OER Project*, https://www.oerproject.com/OER-Materials/OER-Media/HTML-Articles/Origins/Unit7/African-Resistance-to-Colonialism/980L

22 'She Fought for Women's Rights in the French Sudan | Aoua Keita', 15 September 2021, *On the Shoulders of Giants*, https://www.ontheshoulders1.com/the-giants/she-fought-for-womens-rights-in-the-french-sudan-aoua-keita#/.

23 War on Want, 'The Call for Climate Reparations', 15 November 2022, https://waronwant.org/news-analysis/call-climate-reparations.

24 Ibid.

25 Emine Sinmaz, Vikram Dodd and Josh Halliday, 'Thousands of Anti-Racism Protesters Take to Streets Across England to Counter Far-Right Rallies', *The Guardian*, 7 August 2024, https://www.theguardian.com/uk-news/article/2024/aug/07/thousands-of-anti-racism-protesters-take-to-streets-to-counter-far-right-rallies.

26 'In Pictures: Brighton Anti-Racism Protesters Take to Streets', *BBC News*, 8 August 2024, https://www.bbc.co.uk/news/articles/cx2y01r1gk2o.

27 Office for National Statistics, 'Explore 50 Years of International Migration to and from the UK', 1 December 2016, https://www.ons.gov.uk/peoplepopulationandcommunity/populationandmigration/internationalmigration/articles/explore50yearsofinternationalmigrationtoandfromtheuk/2016-12-01.

28 Vince, *Nomad Century*, 138.

Part Three: Land

6. Land

6.1 Our landlessness

1 Malcolm X, *Message to the Grassroots* (1963), published 16 August 2010 by BlackPast, https://www.blackpast.org/african-american-history/speeches-african-american-history/1963-malcolm-x-message-grassroots/.

2 Right to Roam, 'The Right to Roam Is the Right to Reconnect', https://www.righttoroam.org.uk.

3 Michael Sheils McNamee, 'Royal Estates "Receive Millions from Public Bodies and Charities"', 2 November 2024, *BBC News*, https://www.bbc.co.uk/news/articles/cg4l1lzv2nro.

4 'Green Unpleasant Land, Landscape Journal Spring 2022: Whose Landscape Is It?', *Landscape, the journal of the Landscape Institute*, Spring 2022, https://issuu.com/landscape-institute/docs/landscape_journal_whose_landscape_i_a41cbf7b402582/s/14970066.

5 More can be found here: National Trust, 'Addressing Our Histories of Colonialism and Historic Slavery', https://www.nationaltrust.org.uk/who-we-are/research/addressing-our-histories-of-colonialism-and-historic-slavery#rt-the-links-between-wealth-and-slavery-at-places-in-our-care.

6 Ibid.

7 Land in Our Names, 'About Us', https://landinournames.community/about-us.

8 The Rivers Trust, 'Raw Sewage in Our Rivers', https://theriverstrust.org/key-issues/sewage-in-rivers.

9 Lee Bottomley, 'Major Incident Declared after Canal Cyanide Spill', 13 August 2024, *BBC News*, https://www.bbc.co.uk/news/articles/c5yk7nd8r9do.

10 Ibid.

11 Mary Brophy Marcus, 'Indigenous Land Guardianship Around the World', 18 March 2022, *Think Global Health*, https://www.thinkglobalhealth.org/article/indigenous-land-guardianship-around-world.

12 Right to Roam, 'The Right to Roam Is the Right to Reconnect'.

13 Free Siyanda, 'Casey Style Review Demanded', https://freesiyanda.com.

14 Ben McVay, 'Right-Wing Group Unfurl Huge "White Lives Matter" Banner on Mam Tor', 10 July 2020, *Buxton Advertiser*, https://www.buxtonadvertiser.co.uk/news/people/right-wing-

group-unfurl-huge-white-lives-matter-banner-on-mam-tor-2909936.

15 Novara Media, 'Novara Media Reporter Simon Childs "UK Is Fertile Ground for Fascism"', 10 August 2024, TikTok video, https://www.tiktok.com/@novaramedia/video/7401624428716592416.

16 Department for Environment, Food & Rural Affairs, 'Population Statistics for Rural England',14 March 2023, https://www.gov.uk/government/statistics/population-statistics-for-rural-england.

17 https://gal-dem.com/britains-outdoors-green-and-pleasant/.

18 Louisa Adjoa Parker, 'Hidden Histories of the Countryside: Black Lives in South West England', *The Countryside Charity (CPRE)*, https://www.cpre.org.uk/discover/hidden-histories-of-the-countryside-black-lives-in-south-west-england/.

6.2 Nature therapy – A reclamation

1 Jim Robbins, 'Ecopsychology: How Immersion in Nature Benefits Your Health', 9 January 2020, *Yale Environment 360*, https://e360.yale.edu/features/ecopsychology-how-immersion-in-nature-benefits-your-health.

2 Mental Health Foundation, 'Nature: How Connecting with Nature Benefits Our Mental Health', 2021, https://www.mentalhealth.org.uk/our-work/research/nature-how-connecting-nature-benefits-our-mental-health.

3 Emma Travers, 'The grass isn't greener for everyone: why access to green space matters', BMC, 17 September 2020, https://services.thebmc.co.uk/the-grass-isnt-greener-for-everyone-why-access-to-green-space-matters.

4 'Windrush and the Windies - A Celebration of Caribbean British Identity', primary times, June 2023, https://www.primarytimes.

co.uk/news/2023/06/windrush-and-the-windies-a-celebration-of-caribbean-british-identity.

6.3 Some justice, some peace

1 Ta-Nehisi Coates, *Between the World and Me* (London: The Text Publishing Company, 2016), 69.
2 Sheffield Environmental Movement, 'Walk4Health', https://www.semcharity.org.uk/walk4health/.
3 Ibid.

6.4 The space to find out

1 Jennifer Farmer and Zoë Palmer, 'The Dream(ing) Field Lab', *Season for Change*, https://www.seasonforchange.org.uk/commissions/the-dreaming-field-lab/.
2 Ibid.

7. Growing

7.1 The cost of food

1 Errol Schweizer, 'Why Your Groceries Are Still So Expensive', 7 February 2024, *Forbes*, https://www.forbes.com/sites/errolschweizer/2024/02/07/why-your-groceries-are-still-so-expensive/#.
2 'UK Inflation: Why Are Food Prices Rising So Much?', *Sky News*, 16 May 2023, https://news.sky.com/story/uk-inflation-why-are-food-prices-rising-so-much-12860884.
3 'How Much Grain Is Ukraine Exporting and How Is It Leaving the Country?', *BBC News*, 2 April 2024, https://www.bbc.co.uk/news/world-61759692.
4 Schweizer, 'Why Your Groceries Are Still So Expensive'.
5 Ibid.

6 James Davey, 'Britain's Tesco Lifts Profit Outlook after Strong First Half', *Reuters*, 3 October 2024, https://www.reuters.com /business/retail-consumer/britains-tesco-raises-annual-profit-outlook-after-first-half-rise-2024-10-03/.

7 'What You Need to Know About Food Security and Climate Change', *World Bank*, 17 October 2022, https://www.worldbank. org/en/news/feature/2022/10/17/what-you-need-to-know-about-food-security-and-climate-change.

8 Vandana Shiva, *Who Really Feeds the World?: The Failures of Agri-business and the Promise of Agroecology* (Berkeley: North Atlantic Books, 2016), 17.

9 Daniel Macmillen Voskoboynik, 'To Fix the Climate Crisis, We Must Face Up to Our Imperial Past', *openDemocracy*, https:// www.opendemocracy.net/en/opendemocracyuk/to-fix-climate-crisis-we-must-acknowledge-our-imperial-past/.

10 Shiva, *Who Really Feeds the World?*, 21.

11 Ibid., 20.

12 P. Berry, S. Ogilvy and S. Gardner, 'Integrated Farming and Biodiversity, English Nature Research Reports', no. 634 (Peterborough: English Nature, 2004), https://publications.natu-ralengland.org.uk/publication/62023.

13 Jeninaah Hamilton, 'What's Wrong with Monoculture in Farming?', *Youth in Food Systems/ Seeds of Diversity*, 25 February 2022, https://seeds.ca/schoolfoodgardens/whats-wrong-with-monoculture-in-farming/.https://seeds.ca/schoolfood gardens/whats-wrong-with-monoculture-in-farming/.

14 'Biodiversity and Agriculture', *FoodPrint*, 17 February 2021, https://foodprint.org/issues/biodiversity-and-agriculture/.

15 Joe Kobuthi, 'Food Is Power', *Africa Is a Country*, 25 April 2020, https://africasacountry.com/2020/04/food-is-power.

16 'Vandana Shiva and the Hubris of Manipulating Nature' (podcast), *Climate One*, 30 July 2021, https://www.climateone.org/audio/vandana-shiva-and-hubris-manipulating-nature.

17 Stellah Mukhovi and Johanna Jacobi, 'Can Monocultures Be Resilient? Assessment of Buffer Capacity in Two Agroindustrial Cropping Systems in Africa and South America', *Agriculture & Food Security* 11, no. 19 (2022): https://doi.org/10.1186/s40066-022-00356-7.

18 Kobuthi, 'Food Is Power'.

19 African Union, 'Dakar 2: Ghana Country Food and Agriculture Delivery Compact, 2023', https://www.afdb.org/sites/default/files/documents/publications/ghana_country_food_and_agriculture_delivery_compact.pdf.

20 Sam Okyere, 'Priced Out by Imports, Ghana's Farmers Risk Death to Work in Italy', *openDemocracy*, 27 November 2023, https://www.opendemocracy.net/en/beyond-trafficking-and-slavery/priced-out-by-imports-ghanas-farmers-risk-death-to-work-in-italy-migration/.

21 'Senegal Country Commercial Guide', International Trade Administration, U.S. Department of Commerce, 6 May 2024, https://www.trade.gov/country-commercial-guides/senegal-agricultural-sector.

22 Graham Gordon, 'Three Ways Colonialism Contributed to the Breakdown of Our Modern Food System', *CAFOD*, 29 September 2022, https://cafod.org.uk/news/international-news/three-ways-colonialism-contributed-breakdown-food-system.

23 Alice Facchini and Sandra Laville, 'Chilean Villagers Claim British Appetite for Avocados Is Draining Region Dry', *The Guardian*, 17 May 2018, https://www.theguardian.com/environment/2018/may/17/chilean-villagers-claim-british-appetite-for-avocados-is-draining-region-dry.

24 Danwatch, 'Large Avocado Plantations Own Rights to Chile's Water' https://old.danwatch.dk/en/undersogelseskapitel/large-avocado-plantations-own-rights-to-chiles-water/.

7.2 What can grow

1 Office for National Statistics, 'One in Eight British Households Has No Garden', *Office for National Statistics*, 14 May 2020, https://www.ons.gov.uk/economy/environmentalaccounts/articles/oneineightbritishhouseholdshasnogarden/2020-05-14.

2 Tinde Van Andel, 'Talk about the Real Version of "Protective Style"', Know Your Caribbean (@knowyourcaribbean), 22 November 2018, video, https://www.instagram.com/p/BqfclkXB3F-/?utm_source=ig_embed.

3 Shari Rose, 'How Enslaved Africans Braided Rice Seeds Into Their Hair & Changed the World', *Blurred Bylines*, 5 April 2020, https://blurredbylines.com/blog/west-african-slaves-rice-hair-maroon-french-guiana-colonialism/.

4 Friends of the Earth International, 'Food Sovereignty', https://www.foei.org/what-we-do/food-sovereignty/.

5 Amber Murrey, '"Our Stomachs Will Make Themselves Heard": What Sankara Can Teach Us About Food Justice Today', *African Arguments*, 22 April 2016, https://africanarguments.org/2016/04/our-stomachs-will-make-themselves-heard-what-sankara-can-teach-us-about-food-justice-today/.

6 Sang N, 'This Is What President Ibrahim Traoré of Burkina Faso Has Achieved in 2 Years'.

7 Perry Blankson, 'Justice for Thomas Sankara', *Tribune Magazine*, 8 April 2022, https://tribunemag.co.uk/2022/04/thomas-sankara-blaise-camapore-burkina-faso-coup-1987-life-sentence.

8 Patrick Butler, 'More Than A Million UK Residents Live in "Food Deserts", Says Study', *The Guardian*, 12 October 2018, https://

www.theguardian.com/society/2018/oct/12/more-than-a-million-uk-residents-live-in-food-deserts-says-study.

9 Sustainable Food Places, 'Going Hungry in the UK: Our "Food desert" Problem', https://www.sustainablefoodplaces.org/blogs/apr24-food-deserts-in-the-uk/.

10 Leah Penniman, *Farming While Black: Soul Fire Farm's Practical Guide to Liberation on the Land* (White River Junction, VT: Chelsea Green Publishing, 2018), 24.

11 Anna Brones, 'Food Apartheid: The Root of the Problem with America's Groceries', *The Guardian*, 15 May 2018, https://www.theguardian.com/society/2018/may/15/food-apartheid-food-deserts-racism-inequality-america-karen-washington-interview.

12 'Three Sisters (agriculture)', *Wikipedia*, https://en.wikipedia.org/wiki/Three_Sisters_(agriculture).

13 Rich Stockdale 'The Unsettling Truth About UK Land Use', 28 October 2023, https://www.linkedin.com/pulse/unsettling-truth-uk-land-use-rich-stockdale-phd-coqpe/.

14 Khalila Douze, 'Soul Fire Farm: Revolution is Based on Land', 16 August 2022, https://atmos.earth/soul-fire-farm-leah-penniman-afro-indigenous-community-ancestral-wisdom/.

15 Americans Who Tell the Truth, 'Leah Penniman Biography', https://americanswhotellthetruth.org/portraits/leah-penniman/.

16 Ibid.

17 Stephen Heyman, 'Soul Fire Farm's Leah Penniman Explains Why Food Sovereignty Is Central in the Fight For Racial Justice', *Vogue*, 3 July 2020, https://www.vogue.com/article/soul-fire-farm-leah-penniman-why-food-sovereignty-is-central-in-the-fight-for-racial-justice.

18 Soul Fire Farm, 'Farming Practices', https://www.soulfirefarm.org/theland/farmingpractices/.

19 Americans Who Tell the Truth, 'Leah Penniman Biography'.

20 Ibid.

7.3 Herbalism – Community medicine

1 National Institute of Medical Herbalists, 'What Is a Herbalist?', https://nimh.org.uk/_resources/what-is-a-herbalist/.
2 Forest Stewardship Council®, 'Forest Medicines', 21 March 2023, https://fsc.org/en/newscentre/general-news/forest-medicines.
3 Claire Brader, 'Maternal Mortality Rates in the Black Community', *UK Parliament*, 12 December 2023, https://lordslibrary.parliament.uk/maternal-mortality-rates-in-the-black-community/.
4 Julissa James, 'Why Are So Many Turning to Black Herbalists? Their Remedies Are Tailor-Made for 2020', *Los Angeles Times*, 4 December 2020, https://www.latimes.com/lifestyle/story/2020-12-04/black-herbalists-blm-stress-relief-products.
5 Community Apothecary, 'Our Vision: A Web of Healing Medicine Gardens at the Heart of Our Communities', https://www.communityapothecarywf.org/who-we-are.
6 Community Apothecary, 'A Web Of Healing Medicine Gardens At the Heart of Our Communities', https://www.communityapothecarywf.org/.
7 Movement in Thyme, 'Movement & Herbs for the Wellbeing of All', https://www.movementinthyme.com/
8 Movement in Thyme, 'About', https://www.movementinthyme.com/about. Leah speaks of similar reality with their community dropping these things off.
9 The Upsetters, 'The Apothecary Network', https://www.theupsetters.co.uk/the-apothecary-network

7.4 The returning generation

1 Douze, 'Soul Fire Farm'.
2 Penniman, *Farming While Black*, 455.

3 Douze, 'Soul Fire Farm'.
4 Ibid.

Part Four: How We Survive

1 Gwendolyn Brooks, 'Paul Robeson', *Poets.org*, https://poets. org/poem/paul-robeson.

8. Community

8.1 The myth of the monster

1 Liz Mineo, 'A Reading List on Issues of Race: Harvard Faculty Recommend the Writers and Subjects That Promote Context and Understanding', *The Harvard Gazette*, 15 June 2020, https://news.harvard.edu/gazette/story/2020/06/a-reading-list-on-issues-of-race/.

8.2 Loneliness – A climate change story

1 Sarah Johnson, 'WHO Declares Loneliness a "Global Public Health Concern"', *The Guardian*, 16 November 2023, https://www.theguardian.com/global-development/2023/nov/16/who-declares-loneliness-a-global-public-health-concern.
2 Ferdinand Omondi, 'Fast Fashion, Slow Poison: New Report Exposes Toxic Impact of Global Textile Waste in Ghana', *Greenpeace*, 11 September 2024, https://www.greenpeace.org/africa/en/press/56381/fast-fashion-slow-poison-new-report-exposes-toxic-impact-of-global-textile-waste-in-ghana/#:~:text=Air%20Pollution%3A%20Greenpeace%20air%20samples%20from%20public,creation%20of%20'plastic%20beaches'%20along%20the%20coast.

8.3 Moving away from the politics of carelessness

1 We Do Language, 'On Thinking. #bellhooks #blackwriters #booktok #notetoself #fyp', *TikTok*, 22 August 2024, https://vm.tiktok.com/ZGe3xqY5C/.

2 Mia Birdsong, *How We Show Up: Reclaiming Family, Friendship, and Community* (New York: Hachette Go, 2020), 3.

3 Ibid., 4.

4 Mary Kalantzis and Bill Cope, 'Margaret Thatcher: There's No Such Thing as Society', *Works & Days*, https://newlearningonline.com/new-learning/chapter-4/neoliberalism-more-recent-times/margaret-thatcher-theres-no-such-thing-as-society.

5 Sally Weale, 'Youth Services Suffer 70% Funding Cut in Less Than a Decade', *The Guardian*, 20 January 2020, https://www.theguardian.com/society/2020/jan/20/youth-services-suffer-70-funding-cut-in-less-than-a-decade.

6 Polly Toynbee and David Walker, 'The Lost Decade: The Hidden Story of How Austerity Broke Britain', *The Guardian*, 3 March, 2020, https://www.theguardian.com/society/2020/mar/03/lost-decade-hidden-story-how-austerity-broke-britain.

7 Naomi Klein, *Doppelganger: A Trip into the Mirror World* (London: Allen Lane, 2023), 221.

8.4 Knowing each other's names

1 Sherronda J. Brown, 'Left to Hold My Grief Alone: Grieving Platonic Love in a Culture of Romantic Domination. How Do We Mourn Friendship and Community When Romance Inevitably Wins Out?', *Scalawag*, 7 April 2022, https://scalawagmagazine.org/2022/04/grieving-platonic-love/.

2 'Kenmure Street Protests', *Wikipedia*, https://en.wikipedia.org/wiki/Kenmure_Street_protests.

3 Libby Brooks, 'Glasgow Protesters Rejoice as Men Freed after Immigration Van Standoff', *The Guardian*, 13 May 2021, https://www.theguardian.com/uk-news/2021/may/13/glasgow-residents-surround-and-block-immigration-van-from-leaving-street.

4 Nadeem Badshah and agencies, '200 Protesters Block Immigration Officers' Van During Peckham Arrest', *The Guardian*, 11 June 2022, https://www.theguardian.com/uk-news/2022/jun/11/protesters-block-immigration-officers-van-during-peckham-arrest.

5 'SayHerName', *Wikipedia*, https://en.wikipedia.org/wiki/SayHerName.

6 'Detention of Rümeysa Öztürk', *Wikipedia*, last modified 21 April 2025, https://en.wikipedia.org/wiki/Detention_of_R%C3%BCmeysa_%C3%96zt%C3%BCrk.

7 BreakThrough News, 'Happening Now: Thousands of Protestors Call for the Release of Rumeysa Ozturk in Somerville, MA', *TikTok*, https://www.tiktok.com/@btnewsroom/video/7486258043576159534?_r=1&_t=ZN-8vMWeOloPx8.

8 Vilissa Thompson et al., 'Sexual Violence and the Disability Community', *Center for American Progress*, 12 February 2021, https://www.americanprogress.org/article/sexual-violence-disability-community/.

9 Rape Crisis, 'Rape and Sexual Assault Statistics', https://rapecrisis.org.uk/get-informed/statistics-sexual-violence/.

8.5 What Black community means

1 '"Without Community There is No Liberation"– Audre Lorde', *Saffron Press*, 20 July 2024, https://saffronpress.com/without-community-there-is-no-liberation-audre-lorde/.

2 bell hooks, *All About Love: New Visions* (New York: William Morrow Paperbacks, 2016), 130.

Notes

3 *Emergent Strategy*, 10.

4 Ibid., 14.

5 Black Panther Party Alumni Legacy Network, 'Black Panther Party Community Survival Programs 1967–1982', https://bppaln.org/programs.

6 Ibid.

7 Huey P. Newton Foundation, *The Black Panther Party*.

8 Historyin3 (@historyin3), 'Survival Programs', *TikTok*, https://vm.tiktok.com/ZGewXQVJN/.

9 Citizens Advice, 'Credit Union Loans', https://www.citizensadvice.org.uk/debt-and-money/borrowing-money/types-of-borrowing/loans/credit-union-loans/.

10 'How British-Caribbeans Started the First Credit Union in Britain', *Mutual Interest Co-operative*, 11 March 2020, https://www.mutualinterest.coop/2020/03/how-british-caribbeans-started-the-first-credit-union-in-britain.

11 Theodora Lau, 'Lessons for Fintech From Black British Self-Help Money Schemes', *Financial Times*, https://www.ft.com/content/7de2eea4-f030-11e9-bfa4-b25f11f42901.

12 It should be noted that this still going on within the community.

9. Slow Down

1 RIBA Architecture, 'Royal Gold Medal 2024 for Architecture – Winner Professor Lesley Lokko', *YouTube*, April 2024, https://www.youtube.com/watch?v=h4LT1nOlsb8.

9.1 There is no separation

1 ProjectLETS (@projectlets), 'The Earth Is on Fire', *Instagram*, https://www.instagram.com/projectlets/p/C2SeKNZObJi/?img_index=4.

2 https://www.instagram.com/reel/C7hNuFDifJi/.
377

3 The Health Foundation, 'In-work Poverty Trends', 24 July 2024, https://www.health.org.uk/evidence-hub/money-and-resources/poverty/in-work-poverty-trends#:~:text=In%%20per%20cent2D-work%%20per%20cent20poverty%%20per%20cent20trends%%20per%20cent20*%%20per%20cent2063%%20per%20cent%%20per%20cent20of%%20per%20cent20children,13%%20per%20cent%%20per%20cent20in%%20per%20cent202012/13%%20per%20cent20and%%20per%20cent209%%20per%20cent%%20per%20cent20in%%20per%20cent201996/97.

4 Ryan Bradshaw, 'Surprising Working From Home Productivity Statistics', 5 January 2025, *Apollo Technical*, https://www.apollo-technical.com/working-from-home-productivity-statistics/.

5 Raptitude, 'Your Lifestyle Has Already Been Designed', https://www.raptitude.com/2010/07/your-lifestyle-has-already-been-designed/.

6 Matt Swain, 'Bullshit Jobs' (Review of David Graeber, *Bullshit Jobs: The Rise of Pointless Work, and What We Can Do About It* [New York: Simon & Schuster, 2018]), 7 December 2020, https://www.mattswain.com/booknotes/bullshit-jobs.

7 Ibid.

9.3 *The performance of urgency*

1 'How Much Growth Is Required to Achieve Good Lives for All While Reducing Environmental Damage?', *Phys.org*, 25 July 2024, https://phys.org/news/2024-07-growth-required-good-environmental.html.

2 Jason Hickel and Dylan Sullivan, 'How Much Growth Is Required to Achieve Good Lives for All? Insights from Needs-Based Analysis', *World Development Perspectives*, vol. 35, 2024, 100612, https://doi.org/10.1016/j.wdp.2024.100612.

9.4 Slowness for the revolution

1 'Slowness, Ease, and Intention with Latham Thomas, Part 1', Episode 1, *The Emergent Strategy Podcast*, by Emergent Strategy Ideation Institute, https://creators.spotify.com/pod/show/emergentstrategy/episodes/Slowness--Ease--and-Intention-with-Latham-Thomas-Part-1-e1jnmmb/a-a8341i1 31.13.

2 Bayo Akomolafe, *The Times Are Urgent: Let's Slow Down*, https://www.bayoakomolafe.net/post/the-times-are-urgent-lets-slow-down.

3 Oliver Milman and Nina Lakhani, 'Revealed: Wealthy Western Countries Lead in Global Oil and Gas Expansion', *The Guardian*, 24 July 2024, https://www.theguardian.com/environment/article/2024/jul/24/new-oil-gas-emission-data-us-uk.

4 United Nations, 'Only 11 Years Left to Prevent Irreversible Damage from Climate Change, Speakers Warn during General Assembly High-Level Meeting', General Assembly/12131, 28 March 2019, https://press.un.org/en/2019/ga12131.doc.htm.

5 James Baldwin, *The Price of the Ticket: Collected Nonfiction, 1948–1985* (New York: St. Martin's/Marek, 1985), 1989.

9.5 Being safe to one another

1 Tricia Hersey, *Rest Is Resistance: A Manifesto* (New York: Little, Brown Spark, 2022), 26.

9.6 Slowing the means of production

1 Shannon Liao, 'Amazon Warehouse Workers Skip Bathroom Breaks to Keep Their Jobs, Says Report', *The Verge*, 16

April 2018, https://www.theverge.com/2018/4/16/17242766/amazon-warehouse-workers-bathroom-breaks-productivity.

2 Michael Sainato, '"Lack of Respect": Outcry Over Amazon Employee's Death on Warehouse Floor', *The Guardian*, 9 January 2023, https://www.theguardian.com/technology/2023/jan/09/amazon-employee-death-warehouse-floor-colorado.

3 Office for National Statistics, 'Cost of Living Insights: Spending', 13 February 2024, https://www.ons.gov.uk/economy/inflationandpriceindices/articles/costoflivinginsights/spending.

4 Retail Brew, 'Target's Foot Traffic Keeps Falling Amid Boycotts and Slashed DEI Efforts—and It Could Be a Problem That "Isn't Going to Go Away" for the Retailer, Expert Says', 7 April 2025, *Fortune*, https://fortune.com/2025/04/07/target-foot-traffic-keeps-falling-boycotts-slashed-dei-efforts/.

5 CBS Detroit, 'Target Takes Financial Hit Amid Boycott', 9 April 2025, *YouTube* video, https://www.youtube.com/watch?v=5bmRpFr9Pyg.

6 Carrie Arnold, 'Global Life Expectancy Declines for First Time in 30 Years', *Think Global Health*, 11 March 2024, https://www.thinkglobalhealth.org/article/global-life-expectancy-declines-first-time-30-years.

7 Jennifer Dixon DBE, 'Our Health in 2040: Are We Getting Sicker? – With Jeanelle de Gruchy and Kevin Fenton', *The Health Foundation Podcast*, Episode 35, 8 September 2023, https://www.health.org.uk/news-and-comment/podcast/our-health-in-2040-are-we-getting-sicker-with-jeanelle-de-gruchy-and-kevin-fenton.

8 Alyssa Hui-Anderson, 'What Is "Bed Rotting"? Gen Z's Newest Self-Care Trend, Explained', *Health*, updated 20 May 2024, https://www.health.com/what-is-bed-rotting-trend-7561395.

9 Critical Disability Studies Collective (CDSC), 'Terminology', University of Minnesota: Driven to Discover®, https://cdsc. umn.edu/cds/terms#:~:text=Crip%20time%3A%20A%20 concept%20arising.

10 *Telling the Truth*

10.1 *Our forgetfulness in a 'post-truth' era (and the bribe)*

1 Yumi Sakugawa (@yumisakugawa), 'We Must Remember Again and Again Because We Will Forget Again and Again', *Instagram*, https://www.instagram.com/yumisakugawa/p/C4dx CS6vI3A/?img_index=1.

2 Klein, *Doppelganger*, 239.

3 Ian Cobain, Owen Bowcott and Richard Norton-Taylor, 'Britain Destroyed Records of Colonial Crimes', *The Guardian*, 18 April 2012, https://www.theguardian.com/uk/2012/apr/18/ britain-destroyed-records-colonial-crimes.

4 David Child, 'UK Report Denies Systemic Racism, Prompting Angry Backlash', *Al Jazeera*, 31 March 2021, https://www. aljazeera.com/news/2021/3/31/uk-race-report-says-system-not-rigged-against-minorities.

5 'What We Know in the "Marrow of Our Bones"', interview with Clint Smith, *On Being with Krista Tippett*, *The On Being Project*, 2 November 2023, https://onbeing.org/ programs/clint-smith-what-we-know-in-the-marrow-of-our-bones/.

6 Throughline, 'James Baldwin's Shadow', 29 April 2021 (transcript), *NPR*, https://www.npr.org/transcripts/991219491.

7 Ibid.

8 Evan Robinson '23, 'The Joy of an Evening with Angela Davis', *Hamilton College News*, 27 February 2023, https://www.hamilton. edu/news/story/angela-davis-racism-feminism-resistance.

10.2 *Tell the truth, shame the devil*

1 Lancaster University, 'Remember "Covidiots" and the First Protests by "Anti-Vaxxers"? – A New Book on "Viral Language"', 8 November 2023, www.lancaster.ac.uk/linguistics/news/ remember-Covidiots-and-the-first-protests-by-anti-vaxxers-a-new-book-on-viral-language.

2 Adam Gabbatt, 'New York County Signs First Mask Ban into US Law, Sparking Controversy', *The Guardian*, 14 August 2024, www. theguardian.com/us-news/article/2024/aug/14/nassau-county-new-york-mask-ban.

3 Paul McNamara and Anna de Natale, 'Domestic Abuse in the Times of COVID and the Cost-of-Living Crisis', 3 December 2023, *BJGP Life*, https://bjgplife.com/domestic-abuse-in-the-times-of-covid-and-the-cost-of-living-crisis/.

4 Sarah Boden, 'Influenza strain's disappearance, attributed to COVID protocols, alters 2024 flu shot', NPR, 18 October 2024, https://www.npr.org/2024/10/18/nx-s1-5155997/ influenza-strains-disappearance-attributed-to-covid-protocols-alters-2024-flu-shot.

5 Brenda Patoine, 'Child Abuse Actually Decreased During COVID. Here's Why', *Tufts Now*, 14 February 2022, https://now. tufts.edu/2022/02/14/child-abuse-actually-decreased-during-covidCovid-heres-why.

6 Patrick Butler, 'Ill and Disabled People Will Be Made "Invisible" by UK Benefit Cuts, Say Experts', 8 April 2025, *The Guardian*, https:// www.theguardian.com/society/2025/apr/08/ill-disabled-people-uk-benefit-cuts-policy-in-practice.

7 Department for Education, *Academic Year 2023/24: Pupil Absence in Schools in England*, 20 March 2025, https://explore-education-statistics.service.gov.uk/find-statistics/pupil-absence-in-schools-in-england/2023-24.

8 Hannah Croskery, 'Bridging the Gap: Tackling the UK's Literacy Crisis in Our Schools', *Showbie*, 18 September 2024, https://www.showbie.com/bridging-gap-tackling-uks-literacy-crisis-schools/.

9 Tamara Fong, 'Brain fog: Memory and attention after COVID-19', Harvard Health Publishing, 17 March 2022, https://www.health.harvard.edu/blog/brain-fog-memory-and-attention-after-covid-19-202203172707.

10.3 Aggression, 'madness' and ridicule

1 Sara Ahmed, 'Feminist Killjoys (And Other Willful Subjects)', *The Scholar and Feminist Online* 8, no. 3 (Summer 2010), published by The Barnard Center for Research on Women, http://sfonline.barnard.edu/polyphonic/print_ahmed.htm.

2 Travis Alabanza, *None of the Above: Reflections on Life Beyond the Binary* (New York: The Feminist Press at CUNY, 2022), 187.

3 Howard Markel, 'In 1850, Ignaz Semmelweis Saved Lives with Three Words: Wash Your Hands', 15 May 2015, https://www.pbs.org/newshour/health/ignaz-semmelweis-doctor-prescribed-hand-washing/.

4 Ibid.

10.4 We were never meant to survive

1 Audre Lorde, 'My Silences Had Not Protected Me. Your Silence Will Not Protect You', *Goodreads Quotes*, https://www.goodreads.com/quotes/31713-my-silences-had-not-protected-me-your-silence-will-not.

Part Five: How We Thrive

11. Radical Imagination

1 Prentis Hemphill, 'Finding Our Way', Season 3, Episode 2, with guest Bayo Akomolafe, 3 May 2021 (transcript), https://drive.google.com/file/d/1DW1KCbSjv-Gz6OK9RrBpio1uPBVOZnNH/view.

2 Lola Olufemi, *Experiments in Imagining Otherwise* (London: Hajar Press, 28 October 2021), 1.

11.1 Leaving the white man's imagination – Lessons from Ayanda

1 Olga M. Segura, 'TikToker Ayandastood Is Creating Joy Outside the White Imagination', *Bronx Frontlines*, 17 March 2023, https://bronxfrontlines.substack.com/p/tiktoker-ayandastood-is-creating.

11.2 Be unrealistic – Lessons from Walidah

1 'Video of Ursula K. Le Guin delivering her acceptance speech on 19 November 2014', https://www.ursulakleguin.com/nbf-medal.

2 James Baldwin, 'You Think Your Pain and Your Heartbreak Are Unprecedented in the History of the World, but Then You Read', *Goodreads*, https://www.goodreads.com/quotes/5853-you-think-your-pain-and-your-heartbreak-are-unprecedented-in.

3 Angela Washington, 'Afrofuturism in the Stacks', *The Met (The Metropolitan Museum of Art)*, 15 June 2022, https://www.metmuseum.org/perspectives/articles/2022/6/library-afrofuturism.

4 Walidah Imarisha, *Rewriting the Future: Using Science Fiction to Re-Envision Justice*, (blog), 11 February 2015, https://www.walidah.com/blog/2015/2/11/rewriting-the-future-using-science-fiction-to-re-envision-justice.

5 Ibid.

6 Ibid.

7 National Students for Justice in Palestine (@NationalSJP), 'Revolutionary Optimism Is a Wajib', *X* (formerly *Twitter*), May 2024, https://x.com/NationalSJP/status/1794491608538251770.

11.3 Someone who loves me – Ordinary utopias – Lessons from Reagan

1 Ruth Wilson Gilmore and Léopold Lambert, 'Making Abolition Geography in California's Central Valley', *The Funambulist*, 20 December 2018, https://thefunambulist.net/magazine/21-space-activism/interview-making-abolition-geography-california-central-valley-ruth-wilson-gilmore.
2 Reagan Jackson in Natasha Marin (ed.), *Black Imagination: Black Voices on Black Futures* (San Francisco: McSweeney's, 2020), 45.

11.4 God is change – Lessons from Octavia

1 ALOK (Alok V. Menon), 'Octavia Butler says "god is change" what if we were to shift from the unknown as volatility to the unknown . . .', *TikTok* video, 24 August 2023, https://www.tiktok.com/@alokvmenon/video/7270910466862419242.